Concepts, Applications and Emerging Opportunities in Industrial Engineering

Edited by Gary Moynihan

Published in London, United Kingdom

IntechOpen

Supporting open minds since 2005

Concepts, Applications and Emerging Opportunities in Industrial Engineering
http://dx.doi.org/10.5772/intechopen.83281
Edited by Gary Moynihan

Contributors
Mohammad Ammar Alzarrad, Susanto Sudiro, Muhammad T. Hatamleh, Robert Batson, Diogo Ginjo Jantarada, Antonio Grilo, Lokpriya Gaikwad, Vivek Sunnapwar, Mohammad Babul Hasan, Omar Ahmed Raheem Al-Shebeeb, Jose Dinis-Carvalho, Suthep Butdee, Eshetie Berhan, Daniel Kitaw, Siyuan Song, Ibukun Gabriel Awolusi, Gustavo Lopez, Roberto Carlos Valdés Hernández, Adelaida Figueroa Villanueva, Vidblain Amaro Ortega, Gary Moynihan, Yaindrila Barua

First published in London, United Kingdom, 2021 by IntechOpen
IntechOpen is the global imprint of INTECHOPEN LIMITED, registered in England and Wales, registration number: 11086078, 5 Princes Gate Court, London, SW7 2QJ, United Kingdom
Printed in Croatia

British Library Cataloguing-in-Publication Data
A catalogue record for this book is available from the British Library

Additional hard and PDF copies can be obtained from orders@intechopen.com

Concepts, Applications and Emerging Opportunities in Industrial Engineering
Edited by Gary Moynihan
p. cm.
Print ISBN 978-1-83880-935-5
Online ISBN 978-1-83880-936-2
eBook (PDF) ISBN 978-1-83880-937-9

For EU product safety concerns:
IN TECH d.o.o., Prolaz Marije Krucifikse Kozulić 3, 51000 Rijeka, Croatia, info@intechopen.com or visit our website at intechopen.com.

Meet the editor

Gary Moynihan is a Professor and Associate Department Head for Civil, Construction, and Environmental Engineering at the University of Alabama, and was formerly with the Department of Industrial Engineering. He received BS (Chemistry) and MBA (Operations Management) degrees from Rensselaer Polytechnic Institute, and a Ph.D. (Industrial Engineering) from the University of Central Florida. Dr. Moynihan's primary areas of research are project management, operations analysis, information systems development, and the application of industrial engineering techniques. Dr. Moynihan co-founded and served as Assistant Director for the U.S. Department of Energy-funded Alabama Industrial Assessment Center. He has published over 120 articles in books, journals, and conference proceedings, and has five software copyrights. Prior to joining The University of Alabama, Dr. Moynihan held positions in the aerospace, computer, and chemical processing industries.

Contents

Preface

I have been a practicing industrial engineer (IE) for almost forty years. Before that, I had worked in a variety of jobs in manufacturing. My subsequent emerging understanding of IE gave context, meaning, and direction to that which I observed. Operations planning, safety, quality, production control, inventory management, operations research, facility planning, and human factors are traditional IE subdisciplines. From their initial focus in manufacturing, industrial engineering principles, tools, and techniques have spread across a spectrum of application areas. Industrial engineers now work in all areas and all levels of diverse organizations.

This edited book comes at an opportune time. It incorporates new knowledge and expertise in a rapidly changing engineering discipline that is a vital force in a wide range of manufacturing, service, educational, and government organizations. The best practices of 21st century industrial engineering include such concepts as lean systems, sustainability, systems thinking, data analytics, and additive manufacturing, as well as utilization of advanced computer software. These have further expanded industrial engineering's breadth. Each chapter in this book reflects important aspects of these advances.

Dr. Gary P. Moynihan
The University of Alabama,
Tuscaloosa, AL, USA

Section 1

Introduction

Chapter 1

Introductory Chapter: Background and Current Trends in Industrial Engineering

Gary P. Moynihan

1. Introduction

The Institute of Industrial and Systems Engineers' Body of Knowledge [1] formally defines industrial engineering as being "concerned with the design, improvement, and installation of integrated systems of people, materials, information, equipment and energy. It draws upon specialized knowledge and skill in the mathematical, physical, and social sciences with the principles and methods of engineering analysis and design, to specify, predict, and evaluate the results to be obtained from such systems."

While some authors trace the roots of industrial engineering (IE) to much earlier periods, industrial engineering began to define itself during the Industrial Revolution of the 1800s, and particularly the early 1900s. Dr. Batson's chapter, later in this book, includes a fine summary of these historical roots. Frederick Taylor (1911) developed efforts in standardization and specialization, and particularly a focus on workers, their work, and how to effectively manage them [2]. This led to the formalization of such IE sub-disciplines as production planning, scheduling, and inventory control. Nadler [3] further notes that industrial engineering has gone through three broad phases of purpose. The first, building on Taylor's initial work, focused on achieving productivity improvements (mainly efficiency0 in manufacturing plant operations. This early work established the IE profession, and lasted until the late 1920s or early 1930s. The second phase, which lasted until the mid-1980s, extended the efficiency concept with mathematical, statistical and computer-based tools [3]. These included the mathematics of engineering economics, statistics of work measurement and quality control, the modeling and optimization of operations research. As the economy transitioned to an emphasis on the service industry during the 1970s and 1980s, the same techniques and tools of manufacturing-based IE were adapted. These service workers and the systems that they work in require industrial engineering techniques to improve their productivity, just as manufacturing does.

The third phase, according to Nadler [3], shifted from "efficiency to effectiveness and quality, from relatively small systems to large or macro systems". The 20th Century saw the effective application of techniques that progressively subdivide activities to improve operations. More recently, there has been a sharp emphasis on the study of "total systems" in order to optimize operations through the integration of subsystems or parallel systems. For example, the traditional factory-centric perspective in manufacturing application shifted outward to the analysis and improvement of the entire supply chain [4]. Continued expansion and adaptation of IE principles occurred. Whereas, the supply chain once addressed the flow of parts and materials from outside sources to internal company use, supply chain analysis has now been extended to encompass materials, services and information from raw material suppliers, through distribution centers and factories, to the final customer [5]. This establishes a value chain of activities that a firm performs in order to deliver a valuable product for the market.

2. Current status and trends

As implied in a series of articles published by the Institute of Industrial and Systems Engineers (IISE), with the beginning of this third decade of the 21st Century, the field of industrial engineering is experiencing a series of further changes, opportunities and innovations [6]. These emerging trends are built on historical precedent, are not mutually exclusive, and overlap/interweave to a varying degree:

- Continued evolution of applications in the manufacturing and service industries.
- The total systems approach to operational decision-making.
- The use of enhanced data analytics.
- Resulting in the emergence of Industry 4.0.

3. Continued evolution of manufacturing and service applications

As service-centric and factory-centric gains were progressively made, associated improvements in efficiency and effectiveness were obtained. As global markets have emerged, the productivity and competitiveness of companies and nations continue to be a priority. The challenge for IEs is to streamline and better integrate the product or service cycle [5]. The techniques of continuous improvement, six sigma and lean manufacturing, as described in later chapters, support this goal. For example, the concept of lean manufacturing addresses process flow and lead-time then identifies and reduces waste from the process. Six sigma creates value through consistent process output and reducing variation. At the same time, the scope of industrial engineering has expanded to also consider the consequences for safety and sustainability due to increasing public interest, regulatory pressures and corporate social responsibilities [7]. Occupational safety engineering "addresses the origins of workplace accidents, regulations and management practices towards mitigating hazard exposures, preventing harm and reducing liability" [1]. Sustainability refers to practices and efforts that balance the environmental, social and economic needs of current and future generations [7].

4. Total systems approach to operational decision-making

Classical industrial engineering studied the way that people worked in the factories, and the relationship of those workers to their tools and machinery. The focus was on the individual and how to improve the effectiveness of their work. Industrial engineers continue to study how the individual works, but much greater emphasis is being placed upon studying the systems within which the work is performed in order to optimize the performance of the total system by studying the application of knowledge [3]. The need for organizations to develop and implement effective integrated systems has enhanced the profession of industrial engineering. The social sciences provide a source toward which IE looks for information about the behavior of human elements within a system. This is particularly true regarding decision theory and operational analysis [8]. Capturing the desires and judgments of users, stakeholders and customers requires the ability to incorporate such values and model the likelihoods and uncertainties of the alternatives. This role will increase in importance as the decision-making systems of the world continue to grow more complex.

5. Enhanced data analytics

Operations research has long been a specialty area within industrial engineering. It involves the development and application of mathematical models that aim to describe and/or improve real or theoretical systems [1]. This generally involves mathematical optimization to support the decision-making process. For example, the case study provided by Drs. Berhan and Kitaw presents a classic application of linear programming. Subsequent chapters in this section note the development and usage of more sophisticated mathematical models and logic, and their leveraging via computer-based platforms. As data volume, variety and speed of updating increases to support increasingly more complex problems, this linkage of these sophisticated models and computer platforms has evolved into the field of data analytics [9]. Data analytics encompasses such enhanced areas as machine learning concepts, and predictive and prescriptive models.

6. Emergence of industry 4.0

The nexus of these three trends leads to the emergence of what has been termed "Industry 4.0". As noted by Amaba et al. [10], "the terms "Industry 4.0" and "Manufacturing 4.0″ describe the fourth wave of the Industrial Revolution." Each phase was driven by unique technological advances. Industry 1.0 was based on steam power to drive production machines. Industry 2.0 harnessed electricity, mass production and labor division. Industry 3.0 was driven by "computer automation and the use of electronics and IT to further automate production with robotic machines that augmented or replaced operators" [10].

The rise of Industry 4.0 is achieved by integrating digital systems with physical systems (i.e. a cyber physical integration) across the value chain to achieve intelligent manufacturing operations, otherwise known as "the smart factory." Technologies supporting Industry 4.0 include [11]:

- Artificial intelligence and machine learning
- Internet of Things
- Additive manufacturing
- Cloud computing
- Big data analytics

An internet-of-things (IoT) enabled device, broadly defined, is a device connected to the internet, allowing users to access its data and to control its functions remotely [10]. For example, in manufacturing systems, data can be collected from all of the factory workstations to make system operations transparent and enable smart operation decisions to improve various key performance measures [11]. With IoT, a large amount of data from multiple similar subjects/devices/machines are available in real time. The dimension and volume of the data collected is often very large and contains data of different and diverse types. These features set forth the need to rethink many traditional predictive and prescriptive methods to adapt to the unique data features collected, and real-time predictions and decisions often at very high frequencies [12].

7. Summary

For over a century, industrial engineering has arguably been responsible for much of the economic progress in the manufacturing and service industries. Industrial engineering serves people and can lead all workers to improved productivity and to a higher standard of living. The needs for change and continuous improvement are too important, and there is no other profession with the breadth of industrial engineering available to meet them. Of particular importance is the range of directions which the profession can follow. Many threads about the future have been woven into this chapter. Subsequent chapters in this book will illustrate the full range of industrial engineering possibilities.

Author details

Gary P. Moynihan
Department of Civil, Construction and Environmental Engineering, Box 870205, The University of Alabama, Tuscaloosa, AL, 35487-0205, United States

*Address all correspondence to: gmoynihan@eng.ua.edu

References

[1] Institute of Industrial and Systems Engineers. (2019). "Industrial Engineering Body of Knowledge." www. iise.org/Details.aspx? id =43631 (accessed September 22, 2020).

[2] Taylor, F.W. (1911). The Principles of Scientific Management. New York: Harper and Brothers Publishing, 1911.

[3] Nadler, G. (1992) "The Role and Scope of Industrial Engineering". In Handbook of Industrial Engineering (G. Salvendy, ed.) New York: John Wiley & Sons. 3-27.

[4] Jacobs, F. and Chase, R. (2017). Operations and Supply Chain Management, 15th ed. New York: McGraw-Hill Irwin.

[5] Albertson, K. (2020) "Supply chains evolve rapidly to meet changing needs." ISE. 52(5), 34-39

[6] Kelbertson, R. (2020). "2020 and beyond:A future arrives at our doorstep". ISE. 52(1), 6.

[7] Erdil, N.O. (2019). "A focus on sustainability as a core value". ISE. 51(11), 34-39.

[8] Moynihan, G.P. (2018). "Introductory Chapter: Background and Current Trends in Operations Management." In Contemporary Issues and Research in Operations Management (G. Moynihan, ed.) London: InTech Open. 3-8.

[9] Larsen, D. and Chang, V. (2016). "A review and future direction of agile, business intelligenceanalytics and data science". International Journal of Information Management. 36(5), 700-710.

[10] Amaba, B., Clark, V., Cross, D. and McMahon, M. (2020). "AI technology changing the future by enabling Manufacturing 4.0". ISE. 52(4), 28-35.

[11] Wuest, T. (2019). "Smart manufacturing builds opportunities for ISEs" ISE. 51(10), 40-44.

[12] de Oliviera Cordeiro, G., Deschamps, F. and Pinheiro de Lima, E. (2017). "Developing a Big Data analytics project: a case study in the auto industry". Proceedings of the 2017 Industrial and Systems Engineering Conference (K. Coperich, E. Cudney, H. Nembhard, eds.) Atlanta: IISE, 1753-1758.

Section 2

Continued Evolution of Manufacturing and Service Applications

Chapter 2

D Minus 1 Production Scenario: Production Model for Produced Hospital Furniture

Susanto Sudiro

Abstract

Many kinds of production systems are used in medical equipment industries, one of which is through the work-in-process (WIP) buffer control system and feeding material scenarios to assure ability of the process to produce the expected throughput. The production model, known as the D minus 1 production scenario, is used to control production activities at the factory to be carried out using the day minus 1 rule. This rule is a time-based buffer production scenario in 1 day, ending at the finished goods assembly station used as the zero point (D0), from each work-station, pushed for one consecutive day to the beginning of the buffer. With the success of providing WIP buffers on D-1 and D-2 days, the product is certain to be ready on time. Production activities are modeled as Heaviside step function of the various processes involved therein. Production schedule, also production simulation, can be planned through a production dashboard provided for this purpose. Customers demand transformed to an integrated production schedule throughout the production flow, followed by production dispatching and execution. The integrated production schedule includes the supply of raw components, welding, paint, and product assembly to meet on time deliveries.

Keywords: production model, D minus 1, Heaviside step function, WIP buffer, feeding material

1. Introduction

In producing hospital beds or hospital furniture, various manufacturing practices [1–3] and production systems [4–17] are applied in medical equipment industries, one of which is through the work-in-process (WIP) buffer control system and feeding material scenarios so that the process can run normally and produce the expected throughput. The WIP buffer control system and material feed scenario at the factory are known as the D (day) minus 1 production scenario.

Operational production activities are carried out through WIP buffer control and the material stock scenario at the factory using the day minus 1 rule. This rule is a time-based buffer production scenario in 1 day, ending at the finished goods assembly station used as the zero point (D0) from each workstation, pushed for 1 consecutive day to the beginning of the buffer. With the success of providing WIP buffers on D-1 and D-2 days, the product is certain to be ready on time.

Production activities are modeled using a mathematical model of the Heaviside step function [18] of the various production processes involved therein. With these mathematical equations, a production schedule model can be arranged in accordance with the specified production scenarios and it is possible to build a production simulation model in graphic and physical forms through a production dashboard provided for this purpose.

Using this production model, customer demand can be transformed into an integrated production schedule throughout the production flow, followed by production dispatching and execution. The integrated production schedule includes the supply of raw components, welding, paint, and product assembly to meet on time deliveries.

This production model was formed using two dashboards, namely, production planning and scheduling (PPS) dashboard for production planning and production execution management and production information management (PEMPIM) dashboard [19] for production management. By using this dashboard, customer orders can be integrated into the production schedule and the actual production process. In this management model, there is ease of material tracking [20–24] which is used to decide whether an expenditure schedule will be carried out or canceled.

The production dashboard is used to process customer demand data and is used as an input for creating integrated production schedules throughout the production flow, starting from the supply of raw components, welding, paint, and assembly to meet timely deliveries. The production planning schedule produced by the production dashboard is used for scheduling actual production.

Objective of D minus 1 production scenario is to solve manufacturing operation problems, which is implementing the SAP ERP software in a manufacturing company. SAP ERP is an enterprise resource planning software developed by the German company SAP SE. SAP ERP incorporates the key business functions of an organization [25]. The problem is the manufacturing operation in production floor failure to be integrated with the business in company level, and this makes a huge loss for the company [23].

The D minus 1 production scenario has been tested in a factory with actual production operation for various types of products (**Figure 1**). The actual production follows the production schedule in the D minus1 model scenario. Work-in-process buffer (WIP) and material supply scenarios at the plant are controlled using the ease of material tracking facilities. By carrying out actual production, the factory can produce products in the right amount and in time, which adhere to the production schedule so that finished goods can be delivered to customers on time.

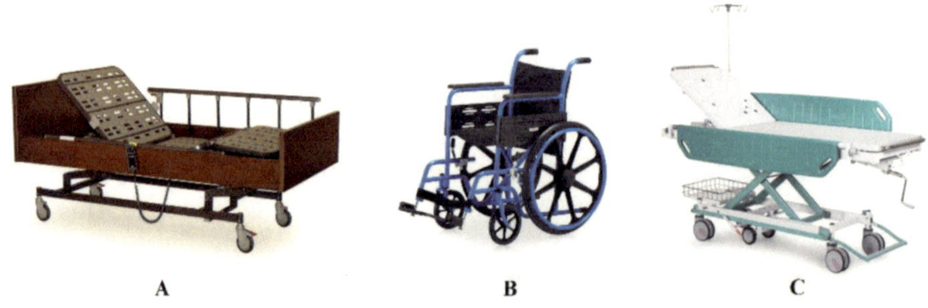

Figure 1.
Many kinds of hospital furniture. (A) Hospital bed. (B) Wheelchairs. (C) Transporting patients.

DOI: http://dx.doi.org/10.5772/intechopen.93691

2. D (day) minus 1 scenario

2.1 Definition of D (day) minus 1

The production model D minus 1 is a production control cycle model that uses N process stages, the planning time horizon is the delivery time of ST (days), the processing time cycle is ST-N + 1 (days) and the process period is daily. Production activities in this case use three stages process, so processing time is ST-2 (days), see **Figure 2**.

The characteristics of this model are:

1. The effective process time for Tp referral is 8 h.

2. In this model production plans can be scheduled for all stations at every stage of the process, this production schedule accommodates the possibility of holidays.

3. The number and type of products that can be scheduled to be produced are various (depending on the requirements on the production floor).

4. From the scheduled production plan, it can be simulated the possibility of delays in the completion of the process so that through this simulation it can be anticipated if a bottleneck is encountered in the process.

2.2 Model material feed scenarios in the factory

The success of the production process is influenced by the role of the material supply scenario in the factory [26, 27], and with a good scenario, it can be ensured that each station receives a timely supply and is easily handled in the production process, because each component is in a standard container rack, trolley, and box (RTB) and in a continuous range of motion so that the operator's movements are merely productive movements to produce added value, while nonproductive movements without producing added value must be kept as low as possible.

The basic principle of supply is to pull in complete quantities. This means that the supply of each process must be complete to form a one-piece flow, and after the material in the WIP buffer is finished, the supply can be withdrawn to the workstation, thus ensuring that the finished goods can be formed because each component is available.

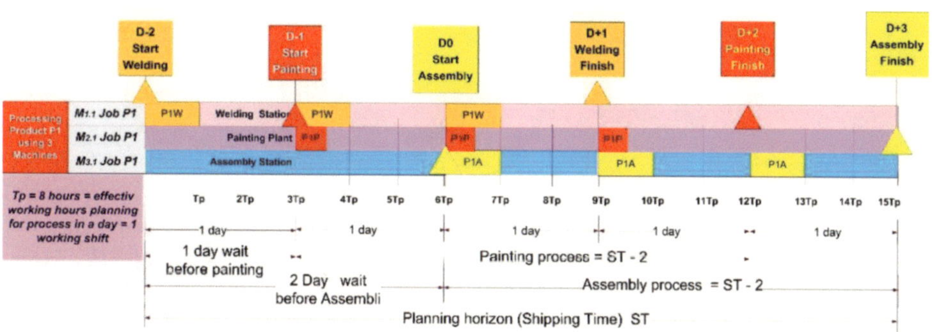

Figure 2.
Planning horizon of the production model D minus 1.

The concept used to fulfill these basic principles is to supply feeding materials to each workstation in the form of lots, batches, or kits. **Figure 3** shows the scenario scheme of the material supply to produce one type of bed. In this case, the production control area is only in the area bounded by the red line; outside of that is the supplier area, and in this case, the supplier is considered capable of meeting the supply requirements as required by his customers.

Starting from the upstream supply, suppliers for metal components in the form of raw components must supply to the welding station in the form of lots, which are placed in a standard RTB for processing at the welding station. The output of the product welding station is sub assay, and the product is placed in a standard RTB in smaller quantities, that is, batch and fed at the paint station.

Furthermore, the results of the paint station are placed in a standard RTB and forwarded to two workstations, namely, the component module station and the final assembly station. The supply of plastic components in batch form is supplied to the component module station and final assembly. The caster wheels are supplied in batches directly to the final assembly station.

The existing supply in the factory is in the form of standard module components and standard component stations. From the station module components are supplied components such as thrusters, side guards or head end foot panels in batch units. Whereas, the supply of standard components is in kit units. The management of this kit form is done in a standard component warehouse.

The production scenario starts with the planned activity at the final assembly station with the production schedule on day 0 (day zero D0) followed by the welding supply time scenario supplied on D-2 (minus 2 day) so that the production schedule is complete, and the supply at the paint station must be available on D-1 day (minus 1 day).

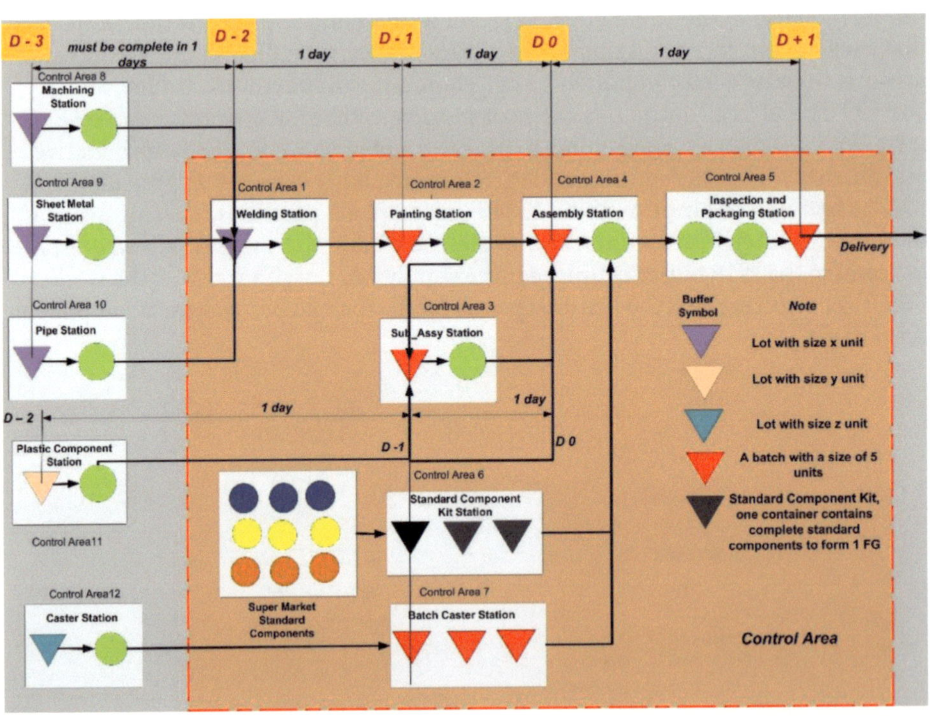

Figure 3.
Material supply scenario.

2.3 WIP buffer model

This section explains the WIP buffer model in the D minus 1 production scenario shown in **Figure 4**, which illustrates the configuration of work-in-process management (WIP) in the form of the production process of making beds for various types of bed products, and the process timing is shown in **Figure 5**.

In **Figure 4**, welding stations there are various stations from M_{11} to M_{1n}, in the paint section there is only one M_2 station, while in the assembly station various stations are available, namely, M_{31} to M_{3n}, each station produces a certain type of product.

All workstations are controlled by the WIP buffer, and the buffer configuration at the factory is from the B_{11} feeder to B_{1n} which is the WIP buffer from the M_{11} to M_{1n} welding station. While the feed in the paint section for the M_2 station is controlled by the WIP buffer continuously at buffer B_{21} to B_{2n}. The output from the paint section continues to hold the WIP buffer from B_{31} to B_{3n} and the buffer is prepared for the M_{31} to M_{3n} assembly station, and the result is the finished product in buffers B_{01} through B_{0n}.

The results of welds from various welding stations cannot be added to the paint section at once, because operations in the paint section are dependent on the hanging effortability of the components to be painted. Each component to be painted is hung on a hanger, and then through the conveyor the components are processed one by one. From the start of hanging until the first component out of the oven takes 90 min, and for a conveyor speed of 2 m/min, a set of products requires time to leave the paint section between 6 min and 12 min, depending on the complexity of the product.

The output from the paint section must be able to feed the M_{31} to M_{3n} assembly stations in accordance with the specified production schedule. If the buffer to the assembly station is done together with the painting station, the M_{31} station needs to wait 1.5 h (the first product waiting period comes out of paint station) plus 60 min (to complete one full provision of one batch of five component sets) while the

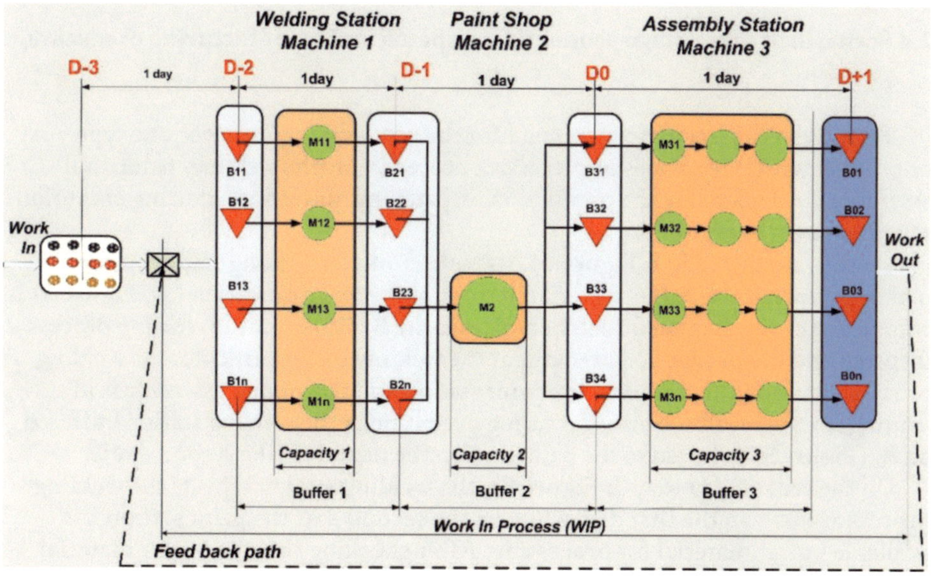

Figure 4.
Production model scenario D minus 1.

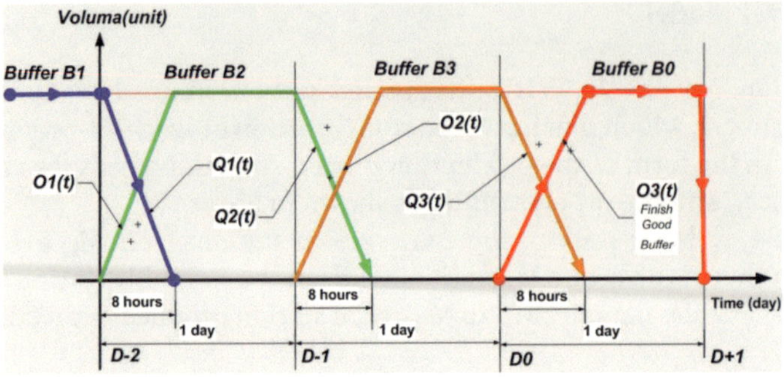

Figure 5.
Timing process in production model scenario D minus 1.

other station waits because there is no buffer. This is a situation where the buffer is not sufficient to supply an assembly station that is designed to operate at a certain capacity.

To overcome this, a buffer scenario is created on day D--2, where sufficient buffers must be available from B_{11} to B_{1n} to be fed to the welding station. Meanwhile, to be fed to the paint part, there must be enough buffers B_{21} to B_{2n} available. Furthermore, to be fed to the assembly station, B_{31} to B_{3n} buffers must also be available in sufficient quantities.

The buffer that needs to be provided is work in. This buffer must be ensured on day D-3 already available, while through the WIP controller the input entered into the system must be controlled, the input is work released, while the throughput is work out provided from the WIP buffer of finished goods B_{01} to B_{0n}.

Through this model, it can be stated that the controlled parameter is WIP in the system, while the manipulated parameter is the upstream buffer in each machine system of the three processing machines. By using the principles of control, of course by controlling WIP through the manipulation of parameters of the production process, it is expected to succeed.

2.4 Scenario of three-stage production dispatch and manufacturing execution process

The production scenario D minus 1 for the production process of one type of product with three stages of successive process, namely, welding, paint, and assembly are depicted in the production dispatching and manufacturing execution scheme shown in **Figure 6**.

On the first day (day 1; **Figure 6**), starting from the welding station, after making sure that the buffer material in front of the welding station (B_1; **Figure 5**) is complete, the production dispatcher releases the B_1 buffer for the welding process dispatch and the production executor at the welding station executes the welding process, while on this first day, the paint station and the second assembly is idle, waiting for the results of manufacturing execution at the welding station buffered on B_2 (**Figure 5**) to be fed to the paint station the next day (the second day).

On the second day (day 2; **Figure 6**), the welding station repeats the welding process as done on the first day, while on the second day, the paint station has available buffer material for processing. After checking that the buffer material in front of the paint station is complete, the production dispatcher releases the buffer for the paint processing dispatch and the paint station production executor executes by carrying out the painting process, while on this second day, the assembly station is still idle, awaiting the results of manufacturing execution at

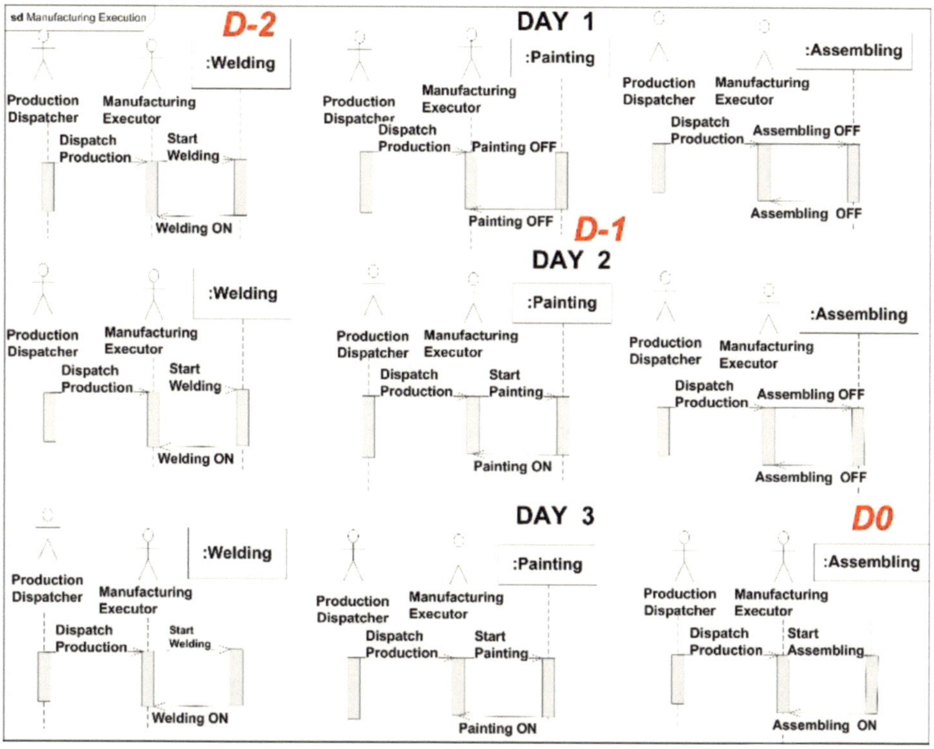

Figure 6.
Production dispatch and manufacturing execution scenarios.

the painting station buffered on B_3 (**Figure 5**) to be fed to the assembly station the next day (the third day).

On the third day (day 3; **Figure 6**), the welding station and the paint station repeat the welding process as done on the second day, while on the third day, the assembly station has a ready component buffer available for assembly. After checking that the finished component buffer in front of the assembly station is complete, the production dispatcher releases the buffer for the dispatch assembly and the production executor at the assembly station carries out the manufacturing execution by carrying out the assembly process to produce finished goods buffered at B_0. On the third day, all three workstations carry out manufacturing executions simultaneously, and this process is repeated in the following days until the specified shipping time is found.

2.5 Planning horizon of the production model D minus 1

The production model scenario D minus 1 use the planning time horizon based on shipping time (ST). This time is described in relation from the beginning of the production process to the completion process (finished goods are sent or stored in warehouses), for all stages of the production process shown previously in **Figure 2**. The work reference is on day 0 (D0), which is the day when the assembly process starts (start assembly), for that on day D-1 (start painting) the painting process must be sure to run, and on day D-2 (start welding) welding process must also be sure to run.

The processing time in the case of three process steps for the welding, paint, and assembly processes is the same as the ST-2 days. While the waiting time to be able to start the painting process is 1 day, while the waiting time for the assembly process is 2 days.

The complete planning horizon for D minus 1 production scenario is illustrated in **Figure 7**. In this figure, the total amount produced during the ST period is expressed in Dt (units), the amount produced daily is Dp (units), and the daily processing time for one shift is Tp = 8 h.

The D minus 1 scenario will be able to be effectively applied to a hospital furniture factory with a daily production plan of more than one type of product (n types of products), because the obstacle is that, at a painting station, one painting station will get feed n buffer components from B_{21} to B_{2n} and must produce finished components which are buffered into n buffers in B_{31} through B_{3n} which will be fed to n assembly stations.

The painting process model is multi-buffer input, single machine, and multi-buffer output, where the buffer output cannot be received immediately even though the input buffer has been fed to the paint station, and the product can only be received at the buffer output after the paint station completes one rotation cycle, and the waiting time for waiting for the first output out of the painting plant is about 90 min.

The results of this paint will later be placed in buffers B_{31} through B_{3n} to be fed to n assembly stations, so if the bait scenario uses the *hot from the oven* method, the waiting time constraints on each assembly station will be a waste, and to overcome this, use the production model D minus 1 with prepared component buffer on minus 1 day before the component is assembled into finished goods. Examples of scenarios for managing four types of products are shown in **Figure 8**.

2.6 Production schedule and example of D minus 1 operation

Production schedule is setup used mathematical Heaviside step function $H(t)$ [18]:

$$H(t) = \begin{cases} 0 \overset{for}{\rightarrow} t \le 0 \\ 1 \overset{for}{\rightarrow} t > 0 \end{cases} \tag{1}$$

Using D_{di} as the stimulus and ST is total planning horizon. General time respond Dp at time t for every number of period k of time period T_d which is time period of a process to finish D_{di} product in a day can be determine as:

$$Dp(t,k) = D_{di}\left(H(t - kT_d) - H(t - kT_d - T_d)\right) \tag{2}$$

k = 1, 2, 3, ST

With Eq. (2) can be determined day-to-day production schedule in each workstations as seen in Eqs. (3)–(5); each is respectively scheduled for welding, painting, and assembly.

$$Dpk(t,k) = D_{di}\left(H(t - (k-1)T_d) - H(t - (k-1)T_d - T_d)\right) \tag{3}$$

$$Dpc(t,k) = D_{di}\left(H(t - kT_d) - H(t - kT_d - Td)\right) \tag{4}$$

$$Dpa(t,k) = D_{di}\left(H(t - (k+1)T_d) - H(t - (k+1)T_d - T_d)\right) \tag{5}$$

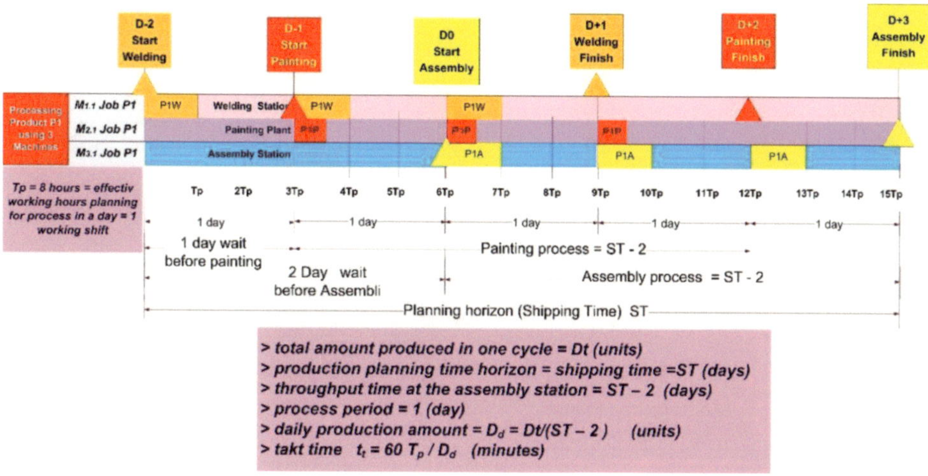

Figure 7.
Scheme of the complete planning horizon of D minus 1 production planning scenario.

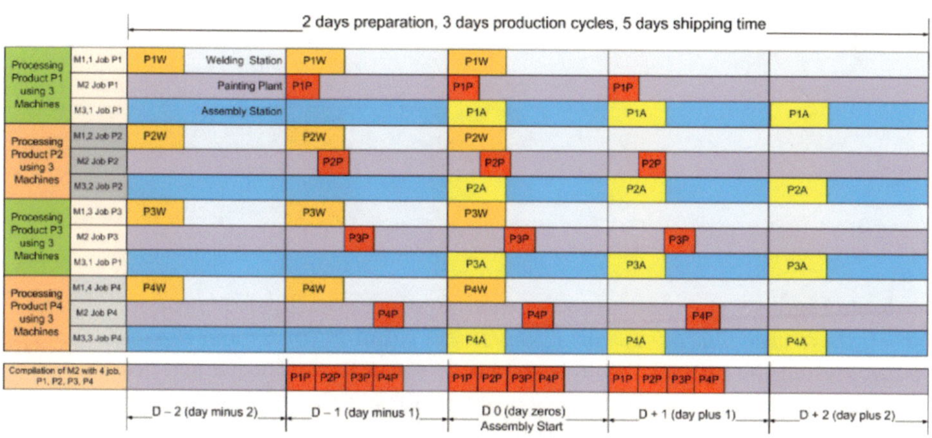

Figure 8.
The planning horizon scenario for the production model D is minus 1 to produce four types of products.

For planning production activity used Eq. (3), Eqs. (4) and (5). Base on those equation be developed application module using Matlab software for setup production schedule. Sample of production schedule is shown in **Table 1**. This is the table of production activity for delivering order of 96 unit products of hospital bed, Supramak 73,006, with daily production of 32 units, with shipping time 5 days; production period is 2÷7 February 2017, in this period at date of 5 February (holidays) production activity is off.

The following will show the scenario D minus 1 of production schedule of **Table 1** to produce products with a total volume of Dt = 96 units, lead time ST = 5 days, daily production plan Dd = Dt/(ST-2) = 96/3 = 32. From this production plan, the production process scenario for the welding cycle time t_{weld} = 18 min as shown in **Figure 9**.

This is a daily production scheme for welding stations using a daily process period of one shift is Tp = 8 h, 1 day is 24 h, so the 24th hour represents the first day, the 48th hour indicates the second day, the 72th hour denotes the third day and so on.

Process	01-Feb	02-Feb	03-Feb	04-Feb	05-Feb	06-Feb	07-Feb
Supply	32	32	32	0	0	0	0
Welding	0	32	32	32	0	0	0
Painting	0	0	32	32	0	32	0
Assembly	0	0	0	32	0	32	32

Table 1.
73,006 Supramak bed production schedule.

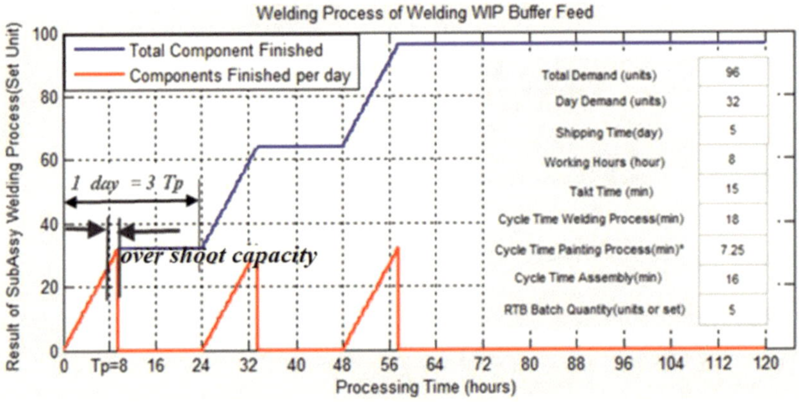

Figure 9.
Example scenario at a welding station. Total production of 96 units, ST = 5 days.

From **Figure 9** at eighth hour, the welding process to finish 32 products should have been completed, but the process was still running (overshoot capacity $T_d > Tp$) to completing the process through extra time (overtime). The occurrence of overtime is due to the available daily production capacity is lower than demand, the indicator is weld cycle time > takt time (18 > 15 min), in this case the overtime that needs to be provided is (18–15) 32 = 96 min.

To avoid overtime in the welding process, in planning the production process, it must be ensured that takt time ≥ cycle time, because takt time indicates the production capacity is associated with workload (number of requests).

The semifinished component buffer for the painting station provided by the welding station the day before was in complete condition, then the component was painted with the process scenario as shown in **Figure 10**.

The complete component supply before leaving the painting station as a finished component, is first circulated throughout the paint station track using a conveyor, so there is a delay in processing time to start the painting process waiting for the finished component to be the first to leave the paint station. In **Figure 10** shown at 24 h.

This delay will reduce the time available for the Tp process to Tp-waiting time, and act as potential to cause a delay in the process of completing the workload (overtime) and this will occur if the number of complete components forming the finished product requires the same cycle time or greater than takt time. But if the paint cycle time is far less than the takt time, then before the process runs out the workload has been completed so that there is still available remaining time and capacity (**Figure 10**), this remaining time can be used to process other types of products.

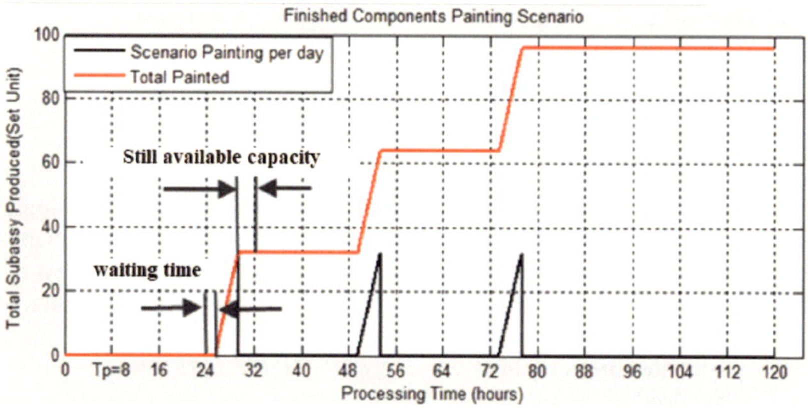

Figure 10.
Example scenario at the paint station. Total production of 96 units, ST = 5 days.

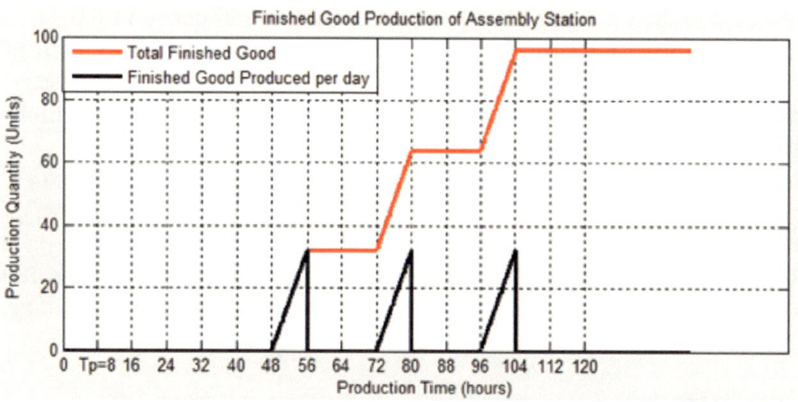

Figure 11.
Example scenario at an assembly station. Total production of 96 units, ST = 5 days.

The finished component buffer for the assembly station provided by the welding station 1 day before is available in complete condition, then the component is assembled with the assembly process scenario as shown in **Figure 11**.

At the assembly station, the process of assembling finished goods can be directly carried out without any waiting time, if the time of the assembly cycle (ta) of the finished product is close to the takt time price then the possibility of overtime can be reduced. **Figure 11** shows that Tp processing time can be utilized maximally, and in this case, overtime does not occur or the process is finished faster than the available processing time.

2.7 Production dashboard

2.7.1 Production planning and scheduling (PPS) dashboard

Operational model of D minus 1 production scenario can be managed using a production dashboard [19]. Starting with production schedules, buffering at each production station and production simulation to total product demand can be demonstrated using this dashboard. Also, daily production activity and calculating the time delay when production cannot be met the target can be demonstrated too. This dashboard besides to simulate production activities based on cycle time and takt time also provides applications to show the response of the production system as a dynamic system.

This dashboard to display production simulation with scenario D minus 1 is called the production planning and scheduling (PPS) dashboard. The PPS dashboard display is shown in **Figure 12**.

Production planning and scheduling can be setup on the dashboard. Before production, event schedule executed the schedule simulated using relevant parameter of production control. If by simulation target of production can be achieved, the scheduled plan decided to be used as production schedule in production floor. But if schedule failure, must be set up new production parameter to control and optimization the process, and with the new parameter once again the production event must be simulated. If the process objective can be fulfilled by this parameter, then the parameter is used in the production schedule.

Control parameter provides in the prototype PPS dashboard, and the parameters used to simulate production process are takt time and cycle time. Cycle time, which is higher than takt time, means capacity of production is lower than the customer demand. For this case the production system is multistage production it is mean all cycle time in each production stage must be equal or lower than takt time. If one of cycle time higher than the takt time, production output will not full fill the customer demand, also there is a bottleneck in one of production stage certainly. If the difference of the takt time and cycle time relatively small, the solution is using extra time in the production floor. But if the difference significantly high the production system must add production time shift.

Control parameter can be directly input to the dashboard use block *Input Parameter* as shown in **Figure 12**. Also using facility provide in the windows dashboard system is with push menu *Input Production Parameter* which is prompt input dialog for input control parameter, the result also displayed in block Input Parameter. For simulate the process must be push each process menu, *Welding*

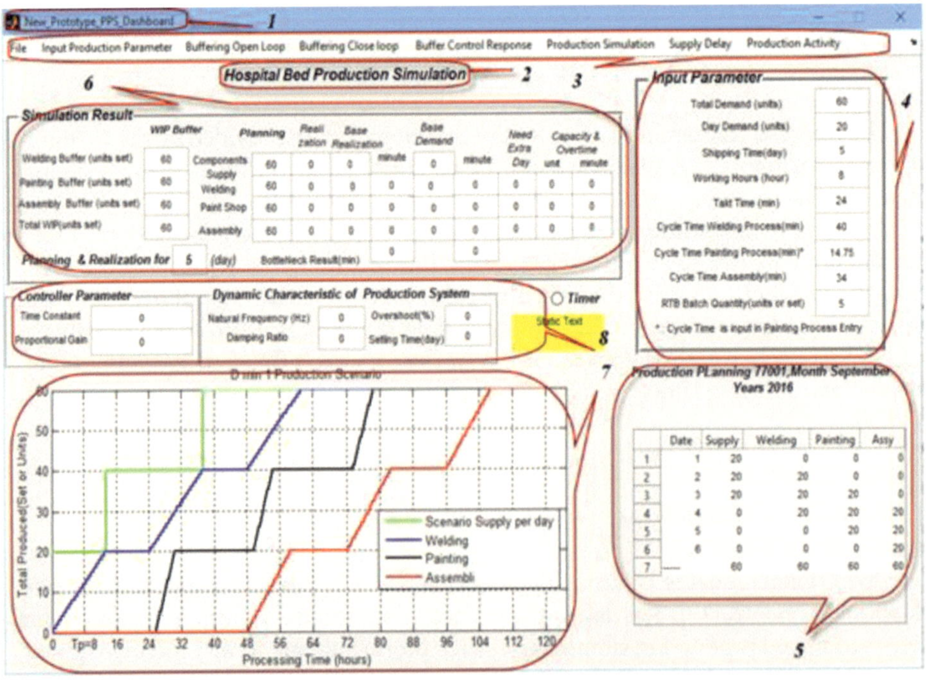

Figure 12.

Production planning and scheduling (PPS) dashboard. The dashboard consists of: 1.Dashboard name, 2. Simulation Title, 3. Dashboard Menu, 4. Input parameters, 5. Product name and production schedule, 6. Simulation Results, 7. Graph Display, and 8. Dynamic system display.

Buffer, *Paint Shop Buffer* or *Assembly Buffer*. To show all process must be push menu *Display result*.

The simulation sample shown in **Figure 12** is simulation of production to produce total customer demand 120 units hospital bed, for shipping time 5 days and day demand is 40 units. Takt time 12 min, cycle time in welding station is 13 min, in paint shop is 7.25 min and in assembly 12 min. Simulation result is shown in *Simulation Result* block.

The dashboard also provide compare between target and realization for any day of production, if there is any difference between target and realization the dashboard will show bottleneck time also need of extra time to finishing the task. If the differences can be accepted use menu in Input parameter to save the production planning.

2.7.2 Production execution management and production information management dashboard

To control production process provide production execution management (PEM) dashboard and production information managements (PIM) dashboard, both dashboard bundle in a single dashboard call as PEMPIM dashboard [19].

The architecture of the dashboard is shown in **Figure 13**, and the windows dashboard is equipped with menu:

1. File menu: For operating file among other open file, save file and close windows.

2. Bill of Materials (BOM) information and status of components supply menu: It is for manufacturing control purposes, consist information of supply components to ensure that supply is complete.

3. Outstanding orders menu: It is for execution purpose of production. This consists of menu for give information of the order status, status finish good in ware house and product delivery to customers.

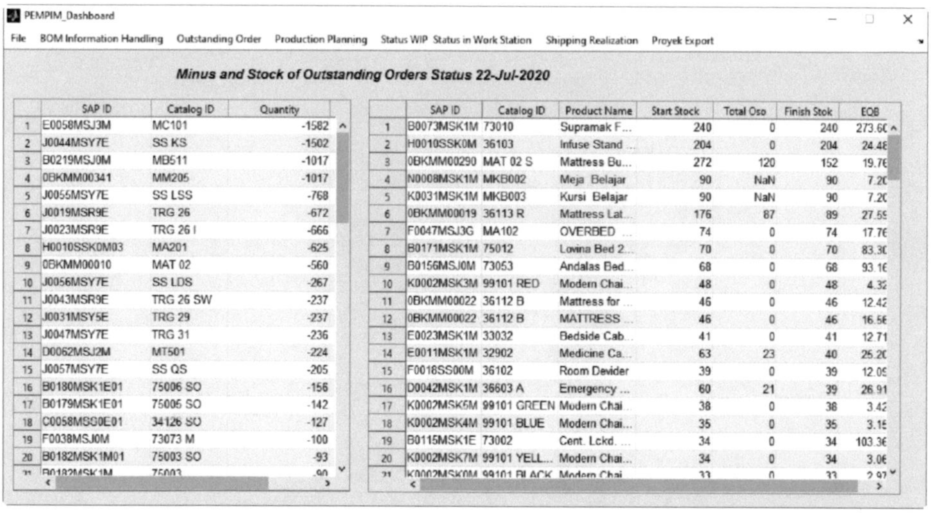

Figure 13.
Production execution management and production information management dashboard.

4. Production WIP menu: It is for manufacturing control purposes, consists information of WIP buffer of supply from internal supplier, welding station WIP buffers, paint shop WIP buffer, and assembly WIP buffer.

5. Target and production realization menu: It is for manufacturing control purposes to show target of production and realization.

6. Machine loading simulation menu: It is additional tool for simulating loading of the machine to forecasting capacity of the workstation.

7. Tool for input components supply menu: It is additional menu for input components supply for special case.

The design of the manufacturing control dashboard is still limited capability as integrated tool for integrating production information from welding station,

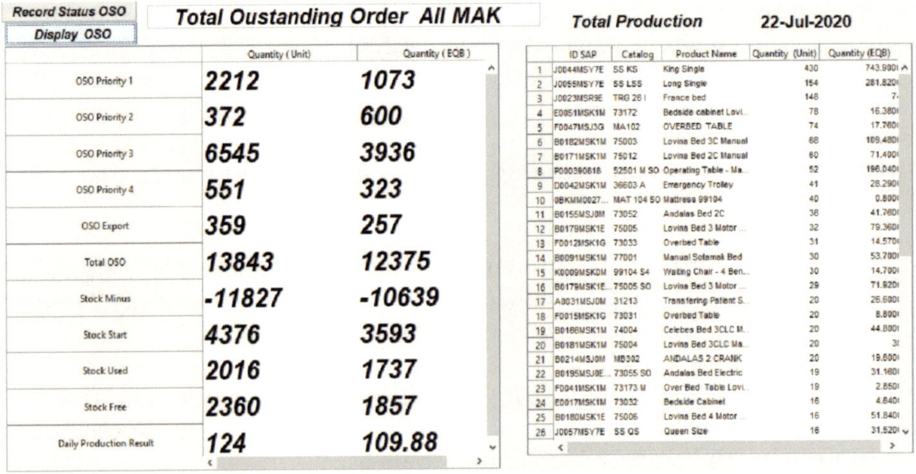

Figure 14.
Snapshot information of outstanding orders status.

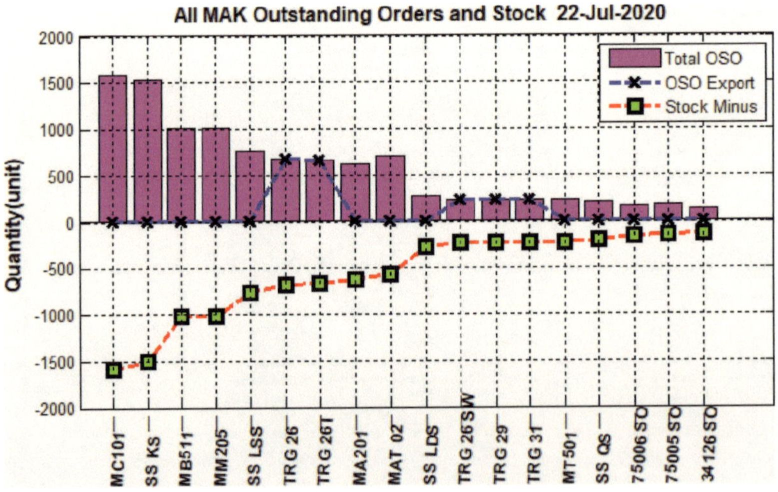

Figure 15.
Snapshot information of outstanding orders and stock.

DOI: http://dx.doi.org/10.5772/intechopen.93691

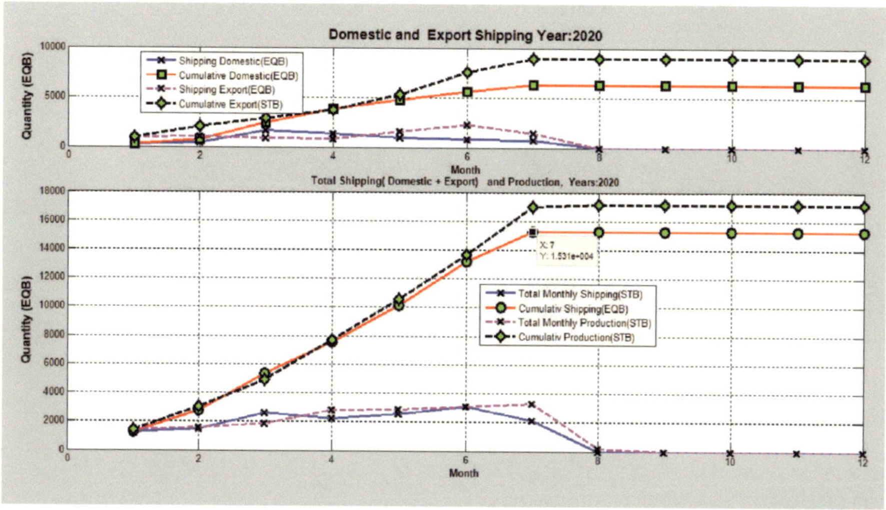

Figure 16.
Snapshot information of production and shipping.

paint shop, assembly also supply of components to be integrated with SAP ERP. The dashboard is a stand-alone dashboard, and not an interface yet to the SAP system.

Examples of output information of the dashboard are: information outstanding orders (**Figure 14**), information of outstanding orders and stock (**Figure 15**), information of production and shipping (**Figure 16**), etc.

3. Verification and validation of the D minus 1 production system

To verify the implementation of this production system, a manufacturing execution system (MES) is used which refers to the ANSI/ISA 95 Part 3 Activity Models of Manufacturing Operations [22, 28]. The manufacturing operations functions are shown in **Figure 17**.

Verification of D minus 1 system is intended to ensure that the functions of manufacturing operations can be operated by providing properly prepared information from the PPS and PEMPIM dashboards, so the production of each workstation normally run and capable to show good production performance.

The verification process begins with processing the product orders into a production schedule and is detailed at all workstations following scenario D minus 1 for the shipping period ST and the detailed schedule must be available at each workstation. This detailed schedule information is then passed on to the production tracking function to ensure that all materials to be used in the process at each workstation are available. For this purpose, the information is provided on the dashboard in the material tracking function.

To ensure that the process can be carried out, the production resources must be available, and for that, the availability of production resources as schedule is verified by using the information on resource requirements at the workstation using the dashboard of available source information.

The verification process is accepted if the system can provide information about the availability of production materials and resources to the dispatcher for each production plan in each workstation according to the production schedule.

To validate the production system is done by validating the dispatcher function and manufacturing execution. This validation process is intended to ensure that the

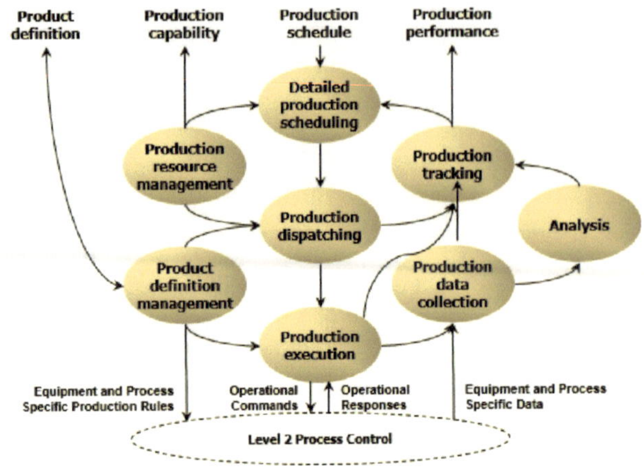

Figure 17.
Manufacturing operation function.

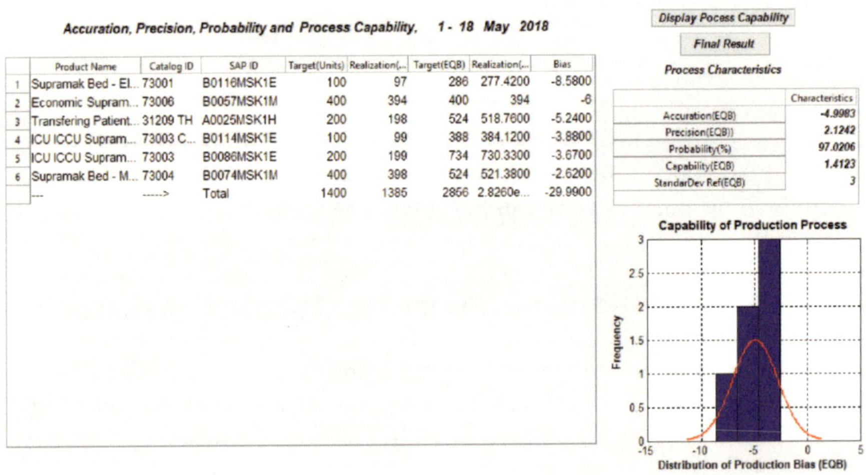

Figure 18.
System validation using capability process.

production process can run and production targets can be achieved. Information on the completeness of resources and the availability of materials must be processed by the dispatcher to determine that the manufacturing execution in each workstation can be carried out.

The process is declared valid if the process accuracy is ±5 equivalent to bed (EQB), precision <= 5 EQB, probability > 90% and process capability > 0.7, it is used as indicators performance for process validation.

The system has been validated on18 May 2018 (**Figure 18**) in production floor, the result all process performance (as process characteristics) indicate fulfilled performance requirement given, it is mean the system is valid.

4. Conclusions

The purpose of D minus 1 production scenario is to produce a management model on the production floor of the assembly plant of hospital furniture through

WIP buffer control and feeding material scenarios to ensure the process runs normally and produces the expected throughput.

The D minus 1 production scenario is an alternative production management model for production scheduling and assembly management from a manufacturing plant of hospital furniture. It also ensures that the production process runs normally and produces a high level of production.

Managing of production activities is done through the control of WIP buffer and material supply scenario in the factory using method D minus 1. This method is a production scenario based on a 1-day buffering period, ending at assembly station as zero point, and then pushed in 1 consecutive day to the beginning of the buffer from each production station. The success of setting buffers at D-1 and D-2 is main factor for the product to be delivered on time.

This production model is built using two dashboards, namely, the PPS dashboard for production planning and the PEMPIM dashboard for production management. By using this dashboard, orders from customers can be integrated into the production schedule and actual production process. Also provided is material tracking facility, which is used to decide whether a production schedule will be implemented or canceled.

The model has been tested in a factory with actual production activity for various types of products. By implementing D minus 1 in the factory, production activity runs normally and is capable to produce products in the right quantity and in a time that complies with the production schedule, so the goods can be delivered on time.

Author details

Susanto Sudiro
Magister of Mechanical Engineering University Pancasila, Jakarta, Indonesia

*Address all correspondence to: susantosudiro@univpancasila.co.id

References

[1] International Organization for Standardization. Medical Devices – Quality Management System – Requirement for Regulatory Purposes. 2016. (ISO 13485)

[2] International Organization for Standardization. Quality Management System - Requirements. 2015. (ISO 9001)

[3] International Organization for Standardization. Medical devices - Application of risk management to medical devices. 2007. (ISO 14971)

[4] Amasaka K. Evolution of Japanese automobile manufacturing strategy using New JIT: Developing QCD studies employing new SCM model. Journal of Applied Mechanical Engineering. 2016;**5**(3):1-14. DOI: 10.4172/2168-9873.1000206

[5] Amasaka K, New JIT. New management technology principle: Surpassing JIT. Procedia Technology. 2014;**16**:1135-1145

[6] Benton WC, Shin H. Manufacturing planning and control: The evolution of MRP and JIT integration. European Journal of Operational Research. 1998;**110**:411-440

[7] Benders J, Riezebos J. Period batch control: Classic, not outdated. Production Planning and Control. 2002;**13**(6):497-506

[8] Berkley BJ. A review of the Kanban production control research literature. Production and Operations Management. 1992;**1**:393-412

[9] Betterton CE, Cox JF III. Expounded drum-buffer-rope flow control in serial lines – A comparative study of simulation models. International Journal of Production Economics. 2009;**117**(1):66-79

[10] Boonlertvanich K. Extended-CONWIP-KANBAN System: Control and Performance Analysis [thesis]. School of Industrial and System Engineering Georgia Institute of Technology; 2006

[11] Bozer YA, McGinnis LF. Kitting versus line stocking: A conceptual framework and a descriptive model. International Journal of Production Economics. 1992;**1992**(28):1-19

[12] Burbidge JL. The new approach to production. Production Engineering. 1961;**40**(12):769-784

[13] Burbidge JL. The use of period batch control (PBC) in the implosive industries. Production Planning and Control. 1994;**5**(1):97-102

[14] Hanson R. In-plant materials supply: Supporting the choice between kitting and continuous supply [thesis]. Sweden: Department of Technology Management and Economics Chalmers University of Technology Gothenburg; 2012

[15] Lage JM, Godinho FM. Variations of the kanban system: Literature review and classification. International Journal of Production Research. 2010;**125**:13-21

[16] Lambrecht M, Segaert A. Buffer stock allocation and assembly type production lines. International Journal of Operations & Production Management. 1990;**10**(2):47-61

[17] Liberopoulos G, Dallery Y. A unified framework for pull control mechanisms in multi-stage manufacturing systems. Annals of Operations Research. 2000;**93**:325-355

[18] Legua MP, Morales I, Sánchez Ruiz LM. The heaviside step function and MATLAB. In: Gervasi O, Murgante B, Laganà A, Taniar D, Mun Y, Gavrilova ML, editors. Computational Science and Its Applications – ICCSA 2008. ICCSA 2008. Lecture Notes in Computer Science. Vol. 5072. Berlin, Heidelberg: Springer; 2008. pp. 1212-1221. DOI: 10.1007/978-3-540-69839-5_93

[19] Sudiro S, Mohd Yusof SR. Design simulation and information dashboard for manufacturing control using systematic approach: A case study of hospital beds production. In: Proceedings of the International Conference on Industrial Engineering and Operations Management; 8-10 March 2016. Kuala Lumpur. Malaysia: IEOM; 2017. pp. 1886-1895

[20] ANSI/ISA-95.00.01-2000. Enterprise-Control System Integration. Part 1: Models and terminology; 2000. ISBN: 1-55617-727-5

[21] ANSI/ISA-95.00.02-2001. Enterprise-Control System Integration. Part 2: Object Model Attributes; 2001. ISBN: 1-55617-773-9

[22] ANSI/ISA-95.00.03-2005. Enterprise-Control System Integration. Part 3: Activity models of manufacturing operations management; 2005. ISBN: 1-55617-955-3

[23] Madapusi A, D'Souzab D. The influence of ERP system implementation on the operational performance of an organization. International Journal of Information Management. 2012;**32**:24-34

[24] Snoeij J. MES product survey. In: LogicaCMG, MESCC – Manufacturing Execution 2006. Arnhem: System Competence Center; 2006

[25] Overview SAP Enhancement Packages - SAP Enhancement Packages - SCN Wiki. Available from: wiki.scn.sap.com [Accessed: 21 March 2017]

[26] Sudiro S, Mohd Yusof SR. Design production schedule and simulation of D minus 1 production scenario using heaviside function and classical control theory: A case study of hospital bed beds production. In: Proceedings of the Asia Pacific Industrial Engineering & Management Systems Conference; 3-6 December 2017. Yogyakarta, Indonesia: APIEMS; 2017. pp. B1-1-B1-6

[27] Sudiro S, Mohd YS. Managing WIP buffer with combination of feeding materials scenario and conventional control theory of single type of hospital bed production. In: Proceeding of Industrial Engineering and Service Science; 1-3 September 2015. Yogyakarta, Indonesia: IESS; 2015. pp. 159-166

[28] Sudiro S, Mohd Yusof SR. Design concept of day minus 1 production model using systems modeling language: A case study of hospital beds. In: Proceedings of the International Conference on Industrial Engineering and Operations Management; 11-13 April 2017. Rabat, Morocco: IEOM; 2017. pp. 304-315

Chapter 3

Continuous Improvement

Dinis Carvalho

Abstract

The principles and concepts originated mainly in Toyota in the 1950s and now presented in models such as Lean Thinking, Shingo Model, and The Toyota Way itself are powerful ideas to bring success to organizations, but the path is not easy. In theory, there are two main features to reach effective and sustainable success, the right vision (True North) and continuous improvement toward that vision. Those two main features are as easy to understand as they are difficult to accomplish. Even when senior leaders have full understanding and belief, the effective implementation on a daily basis is an enormous challenge. This chapter will show how organizations can implement sustainable and stable continuous improvement systems. It will describe the principles that must be followed, the necessary requirements that must be fulfilled, the way the organizations must be structured in teams, the necessary routines, and all the practical steps necessary to reach that sustainable and stable continuous improvement system.

Keywords: continuous improvement, lean thinking, Toyota way, Shingo model

1. Introduction

In the view of many senior industrial engineers and managers, the production philosophy that emerged in Toyota during the 1950s represented a major paradigm shift in the way of thinking about production. The previous milestone had been the mass production paradigm exemplified by the creation of the production and assembly lines. The assembly line concept developed at Ford (curiously another car construction company) revolutionized the industry and society of that time (beginning of the twentieth century). This paradigm produced huge cost reductions and it made possible for people from all over the world to gradually gain access to products that they were unable to have before. There is a less positive side of this way of working, especially in the way operators were seen by the system. The human operator was seen in this system as a purely mechanical resource with the ability to repeat small operations over long periods of time. This repeatability was seen, and is seen still, as a way to achieve high efficiency gains because it is admitted that those who do repeatedly the same operation become masters at doing that operation. With experience, we began to understand the problems that can result from this approach, namely an increase in errors that lead to loss of quality, greater risk of accidents, greater risk of developing musculoskeletal injuries and greater propensity for demotivation at work.

Although Toyota had been using its new production paradigm in its factories since the early 1950s, the first scientific publications on TPS emerged much later. It was not until the 1980s that both Taiichi Ohno and Shigeo Shingo – the two engineers to whom the development of TPS is essentially attributed – published in English about this new paradigm, their concepts and main techniques and practices.

Although Shingo's work (translated by Andrew Dillon), entitled "A Revolution in Manufacturing: The SMED System" (Shingo, 1985), was published earlier, it does not present TPS in general, but rather a concrete methodology, called SMED (Single Minute Exchange of Die), specially designed to reduce equipment setup times. Thus, the first English-language book on TPS was that of Taiichi Ohno, entitled "Toyota Production System: Beyond Large-Scale Production" (Ohno, 1988).

However, it was in 1977 that the first scientific paper on TPS was published in English [1]. This article featured several facets of TPS, including technical aspects such as kanban number calculation (way of controlling production, explained below). However, what aroused the most interest in this publication was one of the two fundamental concepts of TPS. The authors state that TPS is based on two fundamental concepts, the first being the "reduction of the cost of eliminating waste" and the second "treating workers as human beings and with consideration". Although the former is "relatively" easy to understand, the same cannot be said of the second. In fact, none of the concepts are easy to grasp and practice in its fullness, but the second is much more complex. This side of the TPS is also referred by Mike Rother in his book about Toyota Kata [2] as the invisible side of the Toyota approach. This side is also referred by the Shingo model as the social science part of the TPS. It is exactly on this invisible side or social sciences side that the West has struggled to understand and master.

This chapter is dedicated to continuous improvement (CI), and CI is only really effective if the less visible side of the TPS more related to people and culture is seriously taken into account. The continuous improvement (CI) concept is not fully understood by every manager, practitioner, or academic. The three main misconceptions are: (1) CI frequently assumed as being the same as Lean; (2) any improvement is CI, and (3) occasional improvements are CI.

CI is one of the five principles of Lean Thinking and not the same as Lean Thinking. CI is a very important component of Lean but there are other important components as well. Lean without CI is not Lean at all but Lean is more than just CI. There is another word, "Kaizen", which is also used frequently and adds to the confusion. The meaning of *Kaizen* is something similar to continuous improvement but that same word is also used as including the other dimensions that are observed in Toyota factories such as pull production concept, value and waste, visual management, and so on. For many industrial engineers and managers *Kaizen* is similar to *Lean* or TPS itself. CI is only one part of Lean or part of the TPS while the other part is in short a set of principles that define the direction that the CI should follow, as depicted in **Figure 1**.

In contrast, it is wrong to consider any improvement as CI. Any improvement that is not aligned with the defined direction or "true north" cannot be considered as an improvement. Improvements must be increments toward the direction that is defined by the top management people (see **Figure 1**). Let us imagine that the direction is defined as being toward one-piece-flow. The meaning is that the effort should always be in finding solutions to reduce the lot size again and again. Then by some reason a team of engineers promotes an "improvement" that results in a reduction of production cost per unit achieved with bigger lot size and a consequent increase in inventory. Can that "improvement" be considered an improvement? The answer is no because the direction defined was exactly the opposite, the equivalent of pursuing lot size reduction.

Finally, an occasional improvement is not CI. Many articles are published where authors referenced CI which only described a single occasional improvement, either being a setup reduction project or layout improvement project resulting in less transport and motion waste. In reality, these type or actions are important for companies' competitive, no doubt about it, but CI is completely different. CI is

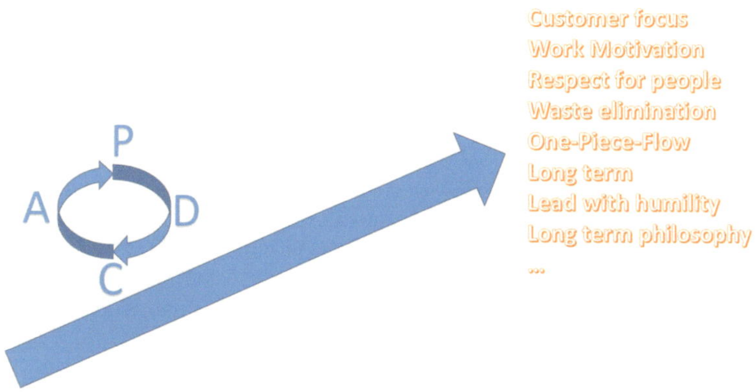

Figure 1.
Continuous improvement direction.

the platform, the systems, the structure, and routines that allow the generation of improvements in a continuous, stable and sustainable way.

2. Reference models

The production management and organization models that are used here as frameworks to CI are all of them inspired from Toyota approach to production or the Toyota model itself. These models will be here described basically in terms of their principles because their principles lead the possible practical solutions developed and implemented in organizations that adopt them.

2.1 Lean thinking

During the 1970s and 1980s, Western companies gradually became aware of what Toyota was doing, especially car manufacturers, because of the impressive results they were achieving. Terms such as *Just In Time* and *Kanban* became very popular as referring the Toyota practices during the 1980s but the word "Lean" as referring the general Toyota approach to production was introduced in 1988 [3] and later popularized by the famous book "The Machine that Changed the World" [4]. The word "lean" attached to words such as "production", and "manufacturing", started to be part of the vocabulary of industrial engineers and managers around the world. Lean as a way of thinking, called "Lean Thinking", was then proposed by [5] and supported by the following five principles:

- Identify value: The idea is that one should try to identify as best as possible what the market interprets as value for the products that the company offers in the market. It is much more important to interpret the value given by the customer to the different characteristics of our product than the value given by the creator of the product itself. This idea is very much linked to popular knowledge "No One Should Be a Judge in his Own Cause". The value should not be given by the cost of the product. Many of the costs that are associated with the product may be associated with characteristics that are not recognized as value by the customer.

- Identifying the value chain: Once identified what is the value of a product from the customer's point of view it is important to identify which processes add value and which processes do not add value and try to eliminate or reduce the latter.

- Create flow: This principle argues that products when they are stopped waiting for the next step in the process represent waste. Ideally, products should always be either transformed in operation (or process) or being transported (as little as possible) to the next operation or process. This principle identifies products that are waiting for the next step, as waste.

- Create customer-pulled production: Implement the production pulled by the customer: ideally means that an operation/process is only performed when the downstream operation/process needs it, with the delivery to the customer being the last of these operations (i.e. the customer "pulls" the production).

- Pursue perfection: This principle is very important and requires major transformations in the structure and culture of organizations. This is where the continuous improvement and the central theme of this chapter is framed.

Lean principles suggest a stronger connection to the physical part of the Toyota approach than to its human side. The term "Lean", in my interpretation, may have been the term chosen by the perception that the researchers had to what they observed in Toyota factories. The reduction of waste ("fat"), especially the reduction of WIP (and the consequent increase in the flow of articles throughout the manufacturing processes) would be easy to observe. This fact may have given the idea of "leanness" as can be interpreted when looking at **Figure 2**.

As it can be seen at the bottom part of **Figure 2**, a smaller amount of WIP is metaphorically represented by a narrower pipe, i.e., it can be said that the production looks "leaner".

Different scholars of Toyota's approach vary greatly in the emphasis they give to the technical aspects and human/culture aspects. If we look at the five Lean principles, we realize that 4 of them are dedicated to technical aspects while only one is dedicated to a mixture of technical aspects and human/culture aspects. It is the principle of continuous improvement or pursuit perfection which is in fact the central theme here in the chapter.

2.2 Shingo model

The Shingo Model was developed in the same period as the Lean Thinking model. The Shingo Model however gives more emphasis to human and cultural

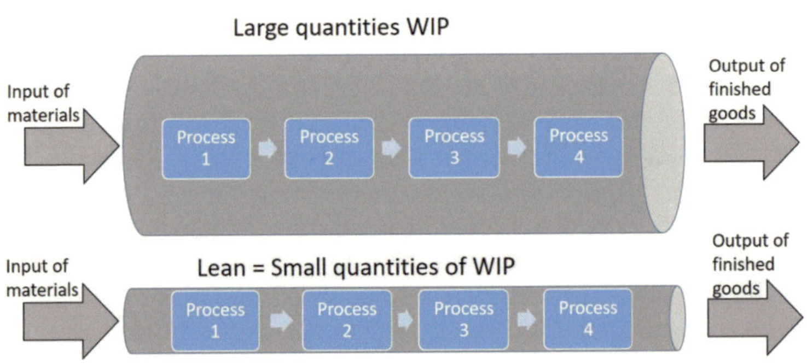

Figure 2.
The physical "leanness" observed with low levels of WIP.

aspects than Lean Thinking. The Shingo Model [6] is the basis of the Shingo prize created by the Shingo Institute in 1988, curiously at the same time that the term "Lean" was being coined by the previously referred MIT group led by James Womack [3].

The Shingo Institute was established in 1988 at Utah State University in the USA following an honorary doctorate awarded to Shigeo Shingo himself. Shigeo Shingo contributed, with Taichi Ohno, with various theories and methodologies for the development of the Toyota Production System. He has written several major books including "A Study of the Toyota Production System: From an Industrial Engineering Viewpoint (Produce What Is Needed, When It's Needed)" first published in English in Japan in 1981 and then published in the US in 1989 [7]. The name of the institute was assigned in his honor. The Shingo Prize was established in 1989 and, in that year the prize was awarded to a single company, located in Bekerley in the USA, called Globe Metallurgical [8]. Since then, it has been awarded to numerous companies that meet the requirements established for this important award of excellence. The Shingo award is based on a set of criteria to assess how "Lean" an organization is, or, if otherwise perhaps more correct, how close to Toyota's way of working an organization is.

The principles that were established in the Shingo Model are grouped into different classes, see **Table 1**. Comparing these principles with the principles of Lean thinking, it is easy to identify differences. Shingo principles are much more comprehensive than Lean principles and probably reflect in more detail the reality of Toyota's philosophy and culture, as they place a lot of emphasis on the less visible aspects of Toyota that are associated with human involvement, vision and culture (**Figure 3**).

The first two principles are clearly dedicated to the side of respect for the individual, referred to by Sugimori et al.[1]. These principles are vital to develop a culture where everyone is motivated to contribute with work and creativity for the benefit of all. These principles are essential to "motivate the elephant" as nicely referred by Chip Heath and Dan Heath in their book entitled "Switch" Heath & Heath (2010). The idea is that even if everyone knows the direction to where their company wants to go, as well as knowing what they are supposed to do in order to go in that direction, no result is expected if motivation does not exist.

Shingo principle	Class
Respect Every Individual	Cultural enablers (People)
Lead with Humility	
Seek Perfection	Continuous improvement (Processes)
Embrace Scientific Thinking	
Focus on Process	
Assure Quality at the Source	
Improve Flow & Pull	
Think Systemically	Enterprise alignment (Purpose)
Create Constancy of Purpose	
Create Value for the Customer	Results (Shareholders)

Table 1.
Shingo principles.

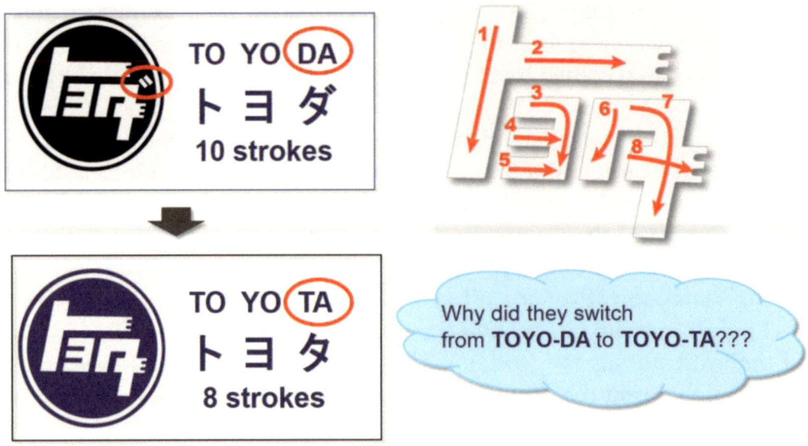

Figure 3.
Toyoda vs. Toyota.

The second class of principles is related to CI. These five principles include almost all technical aspects and behaviors with direct influence in the production process. In this class of continuous improvement, three Lean principles are included, "Seek perfection", "Flow" and "Pull Flow". The principle of "Embrace Scientific Thinking" is also very much related to CI because it brings reliability in problem solving as well as reducing erratic judgment and erratic decision making. Erratic judgment as well as erratic decision making is the classic "jump to conclusions" that happens when managers or supervisors decide very quickly what is the best solution to a problem without using a systematic approach previously defined.

There also two principles related to the enterprise alignment creating the environment where everyone knows where to go (*Create Constancy of Purpose*) and how to act (*Think Systemically*) in order to go toward the vision of the company. The last principle is related to the need to be profitable to sustain the organization.

2.3 Toyota way

The Toyota Way is the title of the book by Jeffrey Liker [9], it is the result of a change of designation that occurred at Toyota in 2001. As the Toyota Production System refers only to the production area, The Toyota Way was adopted to show the intention that the philosophy would extend to all sectors of the organization. As with Toyota, this extension to other areas than just production, has also occurred in various organizations around the world. Thus, with the designation of Lean <anything>, the most varied implementations were appearing, for example, in hospitals (Lean Hospitals), in accounting (Lean Accounting), in administrative environment (Lean Office), in services environment (Lean Services), in academia (Lean Teaching), as well as in many other areas.

This Toyota Way is referred to on the Toyota Europe website [10] as being based on two pillars: Continuous improvement and Respect for People (includes respect and teamwork). It is curious to be close to the concepts that were presented by Sugimori et al. [1] referred to at the beginning of this chapter. Toyota Way is broken down into 14 principles that were published by Jeffrey Liker [9] and which are grouped into the four sections as shown in **Table 2**.

Section	Principle
Section 1 Long-term philosophy	Principle 1. Base your management decisions on a long-term philosophy, even at the expense of short-term financial goals
Section 2 The Right Process Will Produce the Right Results	Principle 2. Create a continuous process flow to bring problems to the surface
	Principle 3. Use "pull" systems to avoid overproduction
	Principle 4. Level out the workload (*Heijunka*)
	Principle 5. Build a culture of stopping to fix problems, to get quality right the first time
	Principle 6. Standardized tasks and processes are the foundation for continuous improvement and employee empowerment
	Principle 7. Use visual control so no problems are hidden
	Principle 8. Use only reliable, thoroughly tested technology that serves your people and processes
Section 3 Add Value to the Organization by Developing Your People	Principle 9. Grow leaders who thoroughly understand the work, live the philosophy, and teach it to others
	Principle 10. Develop exceptional people and teams who follow your company's philosophy
	Principle 11. Respect your extended network of partners and suppliers by challenging them and helping them improve
Section 4 Continuously Solving Root Problems Drives Organizational Learning	Principle 12. Go and see for yourself to thoroughly understand the situation (*Genchi Genbutsu*)
	Principle 13. Make decisions slowly by consensus, thoroughly considering all options; implement decisions rapidly (*Nemawashi*)
	Principle 14. Become a learning organization through relentless reflection (*Hansei*) and continuous improvement (*kaizen*)

Table 2.
Principles of Toyota way.

3. Continuous improvement

Constantly pursuing perfection means never being satisfied with what has already been achieved and always seeking to do better. There is a very interesting story of Toyota, on this subject, which was told by the manager, already retired, Isao Yoshino, who was involved in the installation of Toyota's first factory in the USA. This was the NUMMI (New United Motor Manufacturing, Inc.) project, which resulted from a partnership between Toyota and General Motors and led to the construction of a plant in Fremont, California, in 1984. This car factory was the first experience outside Japan, of implementing Toyota's way of management and culture. Returning to Isao Yoshino, the story he told did not convince me at first, but after a long and informal conversation, I was convinced. This story reveals one of the most fascinating aspects of Toyota's culture, and that is as follows: The word "Toyota" comes from the word "Toyoda", which already existed as a brand (Company Toyoda Automatic Loom Works) and which is actually the name of the family that created that company. In terms of writing, both Latin and Japanese, the words Toyota and Toyoda only differ in tenuous detail. However, it is in Japanese writing that it is important to describe this difference: as can be seen in **Figure 4**, the writing of the two words differs only in two small strokes.

So, 10 strokes are needed to write "Toyoda" while only eight are needed to write "Toyota". Why would they then have changed the name from Toyoda to Toyota, being mentioned that even in this change is present, symbolically, a part of Toyota's culture? The first answer that occurred to me, and that may eventually arise to those who know vaguely the Toyota culture, is that movements are spared when writing the word (fewer traces). Although there is an understandable logic here, this is not the answer. The real reason, explained personally by Isao Yoshino, is that in many cultures, the number 10 symbolizes the top (the maximum) and then, according to this logic, 8 means to be good but still a little below the maximum, which implies that it is necessary to improve to reach the top. This allegory aims to make all the people who work at Toyota understand that you are never good enough that you do not have to improve. As Eiji Toyoda said [11]:

"Being satisfied with the current situation can be the first step towards corruption"

The constant search for perfection (seeking to be 10 assuming that one is always 8) is like a healthy "paranoia", which is based on the fear of letting ourselves relax with the successes of the past and thereby increase the risk of being overtaken by competitors. It is never to allow one to accept that everything is fine, and the truth is that there is always something to improve, there is always some problem unresolved.

On this same topic, Dr. Shigeo Shingo, the co-founder of TPS, said the following: "It is universal truth that those who are not dissatisfied will never make any progress. Yet even if one feels dissatisfaction, it must not be diverted into complaining; it must be actively linked to improvement" [12].

3.1 Continuous improvement versus lean thinking

Continuous improvement is a concept apparently understood by all but actually being able to implement continuous improvement in an organization is a major challenge. Very often, continuous improvement is confused with Lean Philosophy. This subject can be addressed from various perspectives, and may give rise to numerous discussions, but according to one of the publications of the research group responsible for the name "Lean" itself [5], continuous improvement is only one of the five principles of the "Lean Thinking". The other four principles point in the direction in which companies should align improvement efforts, that is, they tell us where continuous improvement should be directed to.

By way of absurd example, suppose that a company assumes as strategic objective to improve the utilization of all its equipment - that is its vision. This is how the company thinks it can lower costs and be more competitive and sustainable (it does not want to have equipment stopped because it wants to monetize its use). According to this objective, all actions that result in increased use of equipment are interpreted as improvements. However, the truth is that these "improvements" may be increasing stocks, worsening quality, increasing response times, and driving the company into bankruptcy. That is why improvement has to be consistent with the direction ("True North"), vision, purpose and values, which are adequate to the survival and sustainability of the company in the long term. Another example: let us imagine that a change introduced on an assembly line resulted in increased productivity. Now let us imagine that this change has also degraded ergonomic aspects or increased the risk of accident, can we still consider it an improvement?

We believe that the most effective and successful direction to align their continuous improvement work, is guided by the principles that underpin the Lean

Philosophy, the Shingo Model, and the Toyota Way. A necessary factor for success occurs when the top decision makers of an organization deeply believe that these principles are the direction to follow. The other major factor is the implementation of the dynamics and routines necessary for the organization to progress in this direction and this is achieved with a continuous improvement system. This is the form of success: define a good direction (Lean Philosophy, Shingo Model and Toyota Way) and start moving in this direction in a systematic and sustainable way (continuous improvement system). In fact, we can only know whether or not there is improvement if we define the direction we want to take. By saying something very obvious, we only know if we move forward if we know which way we want to go.

Many famous stories and phrases allude to this fact. For example, in Alice's story in Wonderland at one point the girl asks the cat: "Where is the exit? It depends, answered the cat. From what? Alice replied. It depends on where you want to go...". Another example is a phrase from the great thinker Seneca: "When one sails without destination, no wind is favorable."

It is essential to define and believe in a path and constantly share it with the whole organization.

3.2 Continuous improvement versus occasional improvements

It is important to clarify the difference between doing improvement and doing continuous improvement. For many, notably Rother [2], continuous improvement is something that happens periodically through technical innovation, projects or improvement events. However, these types of disruptive occasional improvements do not in fact create an effective culture of continuous improvement. As mentioned above, getting an organization to actually have continuous improvement is no easy task and requires very large structural changes. Moreover, continuous improvement can no longer be seen as just a factor of competitive advantage over the competition, but as a survival factor. Companies that do not continuously improve their performance will eventually succumb to their competitors. Continuous improvement is no longer an option and has become the form of survival; the question is: how can each company materialize continuous improvement?

In its fullness CI happens when everyone, every day, implements a small improvement in their area of influence. It is much better for many small improvements to happen continuously than major improvements happening occasionally. When everyone continually contributes to improving their areas aligned with the strategy and the higher purpose of the company, everyone benefits from learning, everyone feels included and everyone grows up with the organization. If the improvements are the result of large projects carried out by external people, there will hardly be a sense of belonging and learning.

The difficulty of implementing and maintaining continuous improvement is enormous, and few companies can do so effectively. It cannot be believed that CI can be achieved and maintained effortlessly and without constant dedication and attention. It is a bit like maintaining equipment or maintaining relationships. It is necessary to maintain constant attention and care. This difficulty is reported in "Toyota Way to Continuous Improvement" [13]. In this work, the authors state that even Toyota has difficulties in maintaining continuous improvement, especially in Toyota factories deployed outside Japan.

3.3 Start with CI

A common first step toward continuous improvement is based on hiring consultants with experience in implement lean tolls to improve processes performance and flows. This path is feasible in the first improvement interventions, but it is not sustainable in the medium and long term. The reason is that, although in the first interventions the consultants can, with little effort, obtain significant performance gains, the following actions will achieve fewer and fewer gains. At some point, the cost inherent in consulting becomes higher than the small gains obtained, and this approach becomes unfeasible. Hiring consultants specialized in implementing improvements according to Lean principles and concepts is very often the approach that companies choose to start adopting continuous improvement practices.

The next step is, for many companies, the development of Lean skills within the organization itself, giving training in this area to a small group of people from top and/or middle management and assigning them the role of drivers for improvement actions, with improvement events or other practices. As the company becomes more mature in terms of continuous improvement, the number of employees who devote part of their time to this type of activities increases more and more, until, in an ideal state, all workforce end up becoming engines of this same continuous improvement. The trend is that more and more companies, and other organizations, start to include continuous improvement activities in the daily routines of all its employees. Companies that do not, will gradually become more and more vulnerable to small crises and market changes.

A very high percentage of companies with continuous improvement systems have suggestion systems. Some of them even integrate rewarding systems, which can contemplate all suggestions or only the best ones. There are cases where the author of the suggestion receives an economic reward proportional to the gain that the company had from the suggestion. This is known as gainsharing. In other cases, all suggestions are rewarded with a fixed premium (cash, gifts, company products, clearances, etc.). In addition, many companies formalize in visual frameworks and internal publications, the recognition, before the whole organization, to this or that employee for their suggestions.

Those who have already gone through the experience of creating suggestion systems quickly learned that the first difficulty is to get employees to start giving suggestions and the second – much more complex – is to follow up on the ideas that begin to enter in the suggestion system. This is another of the topics on which it is necessary to reflect deeply, conscious of unintended consequences, before moving on to implementation. It is notable that Frank Devine's Cathedral/Higher Purpose Model [14], which is highly correlated to success in achieving Shingo awards (25% of all awarded in Europe 2010–2017) explicitly excludes reward as it, as opposed to recognition, did not feature highly enough on the Pareto analysis on which his model relies and is prone to unintended consequences. A frequent temptation is to give incentives for suggestions when you want suggestions to appear. Another is to force a minimum number of suggestions per week for each section of the company. Regarding the first temptation, much could be said but let us mention a very interesting book entitled "Drive" by Daniel Pink [15]. Pink, based on results from many published scientific experiments, suggests, in short, that incentives can be counterproductive when creativity is to be used. As much of what is intended with suggestions is based on people's creativity, it gets a little strange to add something that seems to limit creativity. According to the author, wage incentives are more effective when it comes to increasing productivity in uncreative work, namely repetitive operations without the need for intellectual tearing.

3.4 Resistance to change

It is undeniable that an organization wishing to implement continuous improvement must be prepared for resistance to change. It must be accepted naturally, otherwise it can lead to frustrating situations, and in this sense, it may be important to try to understand.

If it is true that resistance to change is greatly evoked as being present in human nature, it is no less true that the will to change is also present in human nature. These are in fact two seemingly contradictory truths. In a way, all people are willing to change. For example, they are willing to change when they get married or decide to have children, and these changes completely change their lives. Any father knows that having children results in drastic changes in routines, poor quality sleep, constant worries, permanent attention, fear of the unknown, visits to the doctor, and more. Even knowing all the problems that come with having children many people all over the world is deciding to change into it. Many, on the other hand, want to change cars, others from employment, and others from the city, etc. In fact, we are willing to make changes as long as they are the result of our own will. On the other hand, we are not always willing to change when someone wants us to change something in our routines. According to Chip Heath & Dan Heath [16] – for people to change three conditions must be met: (i) clarity of what to change to; (ii) motivation for change and (iii) clarity of the steps needed for that change. With regard to the first condition, it is sometimes not very clear to people what needs to be changed, and this, of course, is undesirable. When you want a team to make a change, you need to make very clear to where or to what they are going to change. With regard to the second condition, it is necessary that the people who will be involved in the change have the necessary motivation to do it. If they do not identify something that is attractive or worthwhile, then these people will not want to change. This is not necessarily a selfish position. People may be motivated for a change that brings benefits to others (and not to themselves) or that is aligned with a belief, a principle, or simply because it is a cause for which they are willing to make the change. Lastly, it is still necessary for people to understand very clearly the steps necessary for a change to occur. Even knowing the reason for the change and having all the motivation to do it, people may not be able to make that change if they do not know what needs to be done.

3.5 Human resources required for CI

In terms of the materialization of CI, ideally the efforts for improvement are part of the normal day-to-day work at all levels of the organization. All people in the organization routinely devote a portion of their daily time to CI tasks. Not spending any time on improvement tasks is bad but spending all the time on improvement tasks is also bad because it is necessary that the work is done for the products to be delivered and the services are provided. Many are the ones who say they do not have time for continuous improvement. On this argument, there is an interesting metaphor:

> *"A lumberjack was cutting down trees uninterrupted with his axe and you could see he was very focused on it. His goal was to cut down as many trees as he could in the time he had. There is however evidence that needs to be met, is that the more cut the less effective the axe to cut. Another man watching approached him and suggested that he should sharpen the axe. The lumberjack replied that he could not sharpen the axe because he did not have time for it."*

This well-known story is referred to by Stephen Covey in his famous book "The 7 Habits of Highly Effective People" [17] In this book, the seventh habit, proposed by the author and which is called "fine tuning the instrument", is directly related to the lumberjack metaphor and also aligned with the principle of continuous improvement.

Returning to the question of what should be the ideal time spent on CI operations, or time spent sharpening the ax, there is a famous metaphor attributed to Abraham Lincoln:

> *"If I had six hours to cut down a tree, I'd spend the first four sharpening the axe"*

In this sentence it is suggested to spend more time on CI tasks than on the tasks of the work itself to add value to products and services. This is not really the practice in organizations. The percentage of time spent on CI is much lower. Although publications that advance with an indicative value are very rare, Mark Hamel & Michael O'Connor argue [18] a value of 3% as a reference for the percentage of human resources that in a company should be dedicated to CI tasks. Interesting value as a reference. However, 3% of the human resources dedicated to CI should be more expensive than 3% of the time of all human resources simply because by assumption the human resources dedicated exclusively to CI, with knowledge of CI and able to promote CI have higher salaries than the average employee.

Considering a round number of 460 minutes that each employee dedicates per day in his work, then 13.8 minutes (3% of 460 minutes) should be used on average for CI tasks. The daily meetings of 5 minutes of the operational teams have a very common practice in companies with organized continuous improvement systems. If we add these 5 minutes daily of all employees with some possible weekly meetings that have longer duration, if we also add the time spent in occasional continuous improvement events and the management and support time in the improvement continues then easily reaches the value of 3% referred.

4. Suggestions systems

An article published in 1999 in the scientific journal Total Quality Management entitled "Stuff the suggestions box" [19] refers to an example of an ancient system of suggestions that existed in Japan at the beginning of the 18th century, created by the eighth Shogun (high-ranking military member) named Yoshimuni Tokugawa. This system of suggestions consisted of a box placed at the entrance of his castle with the following sign: "Make your idea known. Rewards will be given to ideas that are accepted". Rewards were given to people who provided good ideas, but any ideas seen as criticism of the Shogun were often rewarded with beheading.

Modern suggestion systems began to appear in some companies in the early twentieth century. There is reference to suggestion systems in American companies such as Kodak and NCR and Japanese companies such as Kanebo even in the early twentieth century. Later but even before the Second World War, more precisely in the 1930s, companies such as Hitachi, Yasukawa Electric and Origin Electric introduced suggestion systems.

Toyota's suggestion system was introduced in 1951 with the designation "Creative Ideas Suggestion System" and it is described in some detail in the book "40 Years, 20 Million Ideas: Toyota's Suggestion System [20]. It is very interesting to note the importance that Toyota already gave at the end of the first half of the last century to the potential for creativity and innovation of employees. This was so clear that as early as 1953 they formally adopted the slogan "Good Thinking, Good Products" [21]. Both the aspect of respect for creativity and the aspect of putting the customer first were well expressed in that motto.

Operators were led to take a very central role in solving problems that were occurring in the course of their work and also in developing suggestions for introducing improvements. According to some internal Toyota documents I had access to, the attitude of the operators should be as follows:

> *"When we have an idea about a better way to perform the operation, we first clarify the idea with our supervisors and then experiment and evaluate the results. Then we fill out a creative suggestion and submit it for evaluation."*

Each proposal is then verified by a supervisor or group leader, who assesses its usefulness, effectiveness, and originality using a standard scoring system to ensure fairness. Evaluations are carried out almost immediately and a monetary premium is paid about 2 months after the suggestion has been approved.

Another alternative for forwarding suggestions is through quality circles. A quality circle, or quality control circle is a group composed of many people from the same area of work or not, whose goal is to find solutions to solve quality problems or solutions to improve quality. When a good solution is found, that same solution or countermeasure can be applied to all other places facing the same problem. When the members of these quality circles belong to different areas of production, the dissemination of good practices becomes easier. Quality circles were popular in the 1980s but gradually lost their popularity as quality was taken over as part of production. The new type of team that began to exist began to take on purposes other than just quality. This type of team that is often named kaizen teams are the type of teams that would evolve from the initial quality circles. These kaizen teams are a very relevant entity in the structure of most continuous improvement systems.

Although it is assumed that Toyota's suggestion system was inspired by American suggestion systems (such as Ford's) the great effectiveness of Toyota's system is in the emphasis that Toyota has always been given to employees. In texts shared by John Shook[1] in 2009 on suggestion systems it is suggested that one of GM's flaws was clearly neglecting the human aspect of the equation in suggestion systems. Shook argues that GM decided that suggestions that did not bring 20 USD of savings would be refused because the costs to process a suggestion were estimated at that value of 20 USD. The system was focused on the interest of the business and not integrated with the personal dimension. In general terms the conventional suggestion systems existing in America are oriented to encourage GREAT suggestions, to give GREAT prizes, evaluated by LARGE committees with the expectation of GREAT RESULTS. Unlike Toyota encourages small improvements with small prizes. According to the same author, most of the prizes at Toyota are around 10 USD maximum. At Toyota, each employee submits one suggestion per week on average and 98% of the suggestions are implemented. Reflecting on the purpose of Toyota's system the idea is not simply to improve processes, the idea is to improve employee involvement in management and thereby improve processes as a consequence.

In the book "A Study of the Toyota Production System" published 1989 [22] the number of suggestions presented at Toyota by the workers goes from 10.6 suggestions per worker in 1976 to 18.7 suggestions per worker in 1980 and 94% of these suggestions were implemented in that year. The same author suggests that these suggestions have been responsible for the productivity advantage that Toyota's factories have over their competitors. Joakim Ahlstrom [23] found 13 improvements implemented per person per year was a number that very few companies in Europe could achieve. He also mentions that 20 improvements implemented per person per year were the best that was being achieved in the world. The accuracy of these numbers is difficult to guarantee but they are indicators that can be used as a reference.

Yasuda [20] describes the spirit of Toyota's suggestion system in the simple thought: "I want to make my job easier, even if only a little easier". It also says: "If people remain alert to detect problems, in an environment created that facilitates

[1] John Shook is one of the greatest figures in the Lean movement. He is the author of several books on lean as the famous "learning to see" dedicated to VSM (Value Stream Mapping). John Shook was involved in the first experience of applying TPS in America, the NUMMI project. It was a car factory that is born from a consortium between Toyota and GM.

their identification, the generation of creative ideas will not be exhausted". The idea of making very frequent small improvements seems counterintuitive in the culture of most of our companies. In many companies we are always looking for a very good solution that makes us have significant performance gains and we do not give much credit to insignificant or sometimes intangible gains but that, in the long run, by their frequency, are extremely powerful. This is a key idea for the sustainability of continuous improvement.

In 'Human Resource Development in Toyota Culture' [24] the authors mention that the relationship of trust created between the employee and the company or institution is extremely important and perhaps that is the reason why 90,000 suggestions per year are presented by Toyota employees. It is also curious to note that, according to the same authors, Toyota intentionally does not give emphasis to cash rewards. This is consistent with Devine's work stressing recognition rather than reward [14]. Generating ideas is supposed to be part of the work and as such does not reward a person for having done only his work. The ideas generated are part of the relationship of trust between the parties and this relationship of trust is the currency of exchange between employer and employee. Employees individually can and should place suggestions in suggestion boxes placed in specified places. They are encouraged to write them or ask for help from more experienced engineering personnel to write them if necessary but can also give suggestions verbally so that there is no excuse for suggestions not to be given.

Toyota handles the suggestion system in a very similar way to how the system was originally implemented many years ago. There is a committee that evaluates all the suggestions of the workers. Whenever a worker has an improvement idea in his work, he first talks about his idea with his direct boss, as a team leader, and tries to implement it as a test to find out how the idea works. If they find it would work, the worker will put the idea in a specific predefined format, with the support of the boss, if necessary, and formally send the idea to the committee. The committee then assesses the suggestion very quickly and decides whether it is implemented or not and what the reward should be. However, it is important to emphasize that the main objective of the system is NOT to give money to the employee who sends the suggestion. The main objective of donating a small amount of money is to APPRECIATE the worker's initiative and encourage him to continue the approach. Therefore, the main objective of the system is not to compensate the worker with the money, but to encourage his attitude.

5. Main features of effective CI systems

To be effective a CI system needs top managers to master not only the technical management skills side but also the behavior and cultural side. Mike Rother [2] refers to the visible and the invisible side. The visible part has been more easily copied from Toyota many times without success because the invisible part is difficult to copy and is often overlooked. This type of division is also referred to in other forms such as the Shingo Model, dividing the KPI (Key Performance Indicators) and KBI (Key Behavioral Indicators) [25].

This invisible side of organizations or the side of behavior and culture is as present and vital as it is neglected. Edgar Schein argues: "The only thing leaders do of real importance is create and manage culture. If you don't manage the culture, the culture manages you, and you may not have the slightest idea of the extent to which it is happening."

Leaders in the organization must define the long-term vision trying to include a higher purpose that may create among all workers a sense of being part of something big. Something that makes them proud of being part of such organization. Leader and managers must also fully understand principles such as "respect for people" and "lead with humility". These principles must be more than just understood, rather they must be really living in their hearts. Devine's 'Rapid Mass Engagement Process [reference Conference Book?] amplifies this effect when the employees themselves create their own CI culture, and, sometimes, their own Higher Purpose by consensus with their senior leaders. In the face of a problem managers must behave in a systematic way starting with a "not blame and not judge" attitude. Most problems occurred because the systems allow them to occur and blaming on workers is the wrong approach to start with. The idea is to try to find the root cause and eliminate it. Manager must welcome problems and encourage workers revealed them since they promote improvement.

5.1 Operational teams

Another important feature of CI is teamwork. A common feature in effective CI systems is the presence of teams everywhere in the organization. Workers are organized in teams and managers gradually empower those teams with management responsibilities such as performance measurement and monitoring, scheduling their work, problem solving, and routine meetings. Transparency is an important tool to the benefit of all. Visual management must be used by teams to show and update daily most important aspect of their work.

These operational teams, also sometimes called "natural teams", are teams that normally share the same workspace and the same tasks, or complementary tasks. The term "natural" is used because in fact these teams already exist in a natural way even if they have no formal routines, no goals, and no team spirit. These teams are the essential units of CI systems as they are largely responsible for adding value to products and as such where CI is first most important. When we hear the term "operational teams" we often associate with teams operating in the production units, but the operational teams extend to all levels of the organization. These teams can be teams with administrative tasks, development tasks, sales or even management tasks.

The size of a team is a major factor as their performance can vary with their size. A good rule is to try to create small teams. On this team size issue, Sutherland [26] suggests that teams of 3–7 people needed 25% less effort than 9–12 people to do the same job. The size of the team is often determined and/or forced by constraints of context and typology of personality and relationship between team members. There are teams of 3 that work very well but others there are the same number but that do not work at all. Whether or not a strong team leader exists is also a key factor in the team's performance and ideal size.

5.2 Teams meetings

The meetings of the operational teams are events of enormous importance for the sustainability of continuous improvement. The frequency of these meetings is commonly proposed as one meeting per day. In the Kaizen Institute CI model, it is even called Daily Kaizen. Typically the meeting follow a standard agenda lasting 5 minutes and in general terms the objectives of these meetings are: to preserve and improve the cohesion and motivation of the team, to become aware of the state and evolution of performance indicators, to plan the work of the day, to monitor ongoing improvement actions and if necessary to report occurrences. For these meetings to be effective it is necessary for teams to be trained and monitored very closely in the first few weeks. It is necessary that someone with experience accompanies the first meetings giving suggestions and feedback on the content of the agenda and how to conduct the meeting. At first these meetings tend to take more than 5 minutes and some issues remain un-addressed but as the team gains experience it becomes more effective in managing time and eventually manages to do the whole meeting in about 5 minutes in most cases.

It is important to repeat that the meeting cannot include any kind of blaming or judgment of team members. The spirit of the meeting should always be positive and motivating. It should always be sought to encourage team members to try to identify something that may be preventing them from doing their job better, with less effort and less risk.

In some organizations, workers are trained in coaching and in constructive feedback in order to help creating the perfect atmosphere to approach problems, improvements, and improve motivation and work satisfaction. The well-being of team members should always be valued because respect for the human person has an unquestionable intrinsic value and also because people who feel good can produce better, with more quality and with more safety. It is important to remember what was said in the first article published in English about the Toyota Production System (TPS). In this article it is said that one of the two basic concepts of TPS is "treating workers as being human and with consideration" [1]. It is important that the whole organization breathes awareness of this concept and that it creates standards, routines and practices that materialize this concept and that gradually becomes the culture of the organization.

5.3 Project teams

The effort for continuous improvement must be widespread throughout the organization. That is true, however, although everyone in the organization must be involved in the constant search for improvement, it is common the need to create temporary teams that aim at significant improvements in the process, also called disruptive changes.

Disruptive project teams are teams designed to introduce disruptive process improvements to clearly improve performance levels and typically lead to changing working standards. These new standards are then followed by natural teams in order to stabilize them. In some companies the main aim of operational teams is to stabilize standards and never improve them. These projects are projects for significant changes in the process, relate to striking/disruptive improvements that result in layout changes, task change, line resizing, technology changes in the process, reduction of product change times, etc. Unlike natural teams that remain stable for a long time these teams are formed when the project begins and is extinguished when the project ends.

These disruptive Kaizen events typically are held for 1 week every 3 months. Kaizen events should start with a well-defined problem, described in a clear and short sentence, to enhance the effectiveness of the event and should also be seen as an investment for their very effective role in training employees [27]. Bastos & Sharman [28] suggest that for there to be significant improvements, this level of action called the disruptive improvements, together with the CI performed by natural teams, has a multiplier effect on the improvement of processes and consequently on the performance and competitiveness of an organization. The main message is that the combination of the two types of efforts for improvement (Continuous and Disruptive Improvement) result much more effectively than just the CI. In contrast Mike Rother [2] argues:

> *"... Toyota considers the improvement capability of all the people in an organization the "strength" of a company. From this perspective, then, it is better for an organization's adaptiveness, competitiveness, and survival to have a large group*

> *of people systematically, methodically, making many small steps of improvement every day rather than a small group doing periodic big projects and events."*

Despite that, Mike Rother also argues:

> *"... If your business philosophy is to improve, then periodic improvement projects or kaizen workshops are okay but not enough..."*

5.4 Coaching

As in everything in life the CI systems needs be maintained and improved to stay alive. Some kind of structure is necessary to maintain, coach and guide natural teams and promote improvement projects. In the Toyota Kata (TK) model presented by Mike Rother [2] represents how the Toyota CI system works. The role of maintaining the natural teams' moral and focus on seeking perfection is assigned to a coach. The coach's role is very relevant in the TK model since it is up to the Coach to play the main role of maintaining the Toyota culture, maintaining the team's motivation and guiding it to the identification and implementation of improvement actions that ensure learning, clear knowledge of the process and evolution toward challenges and vision. The structure that ensures the sustainability of continuous improvement is represented in **Figure 4** where the coaching structure can be represented. Each team leader meets with his Coach once a day (whenever possible) in front of his team's board for a follow-up session of the actions and evolution of the team always having as support the data that is visually presented on the team board. The team leader's daily sessions with the coach are occasionally accompanied by a second-level coach who will help the first-level coach evolve his coaching skills.

When a vacancy for 2nd level coach emerges, the 1st level coach, who is more advanced in terms of coaching skills, will be selected. The vacancy left by the first-level coach will have to be filled by a team leader with more coaching experience and skills. This vacancy in turn will be occupied by the team worker with more leadership skills. In this way the coaches were workers who absorbed Toyota's culture and assumed the responsibility of maintaining it and passing it on to the new workers.

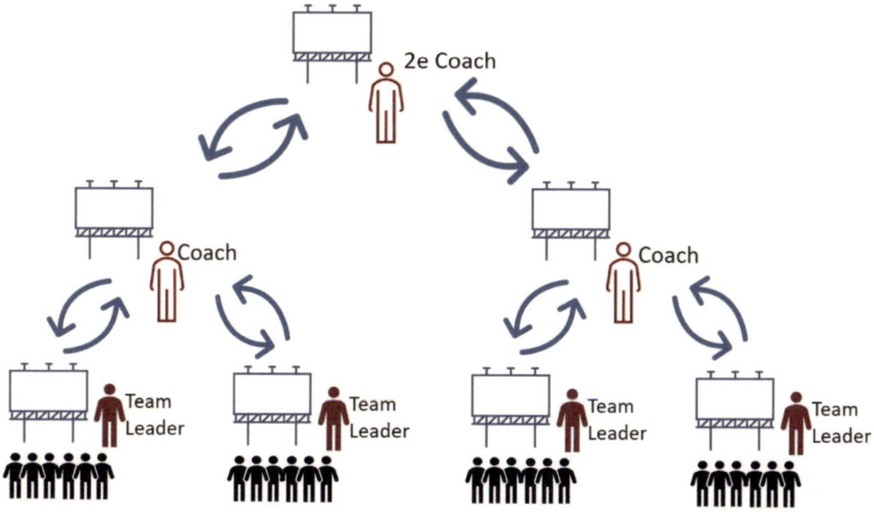

Figure 4.
Coaching structure of the TK.

6. Conclusions

Continuous improvement is a key feature to include in the culture of organizations that want to play a major role in people and society. The CI culture with the right direction and higher purpose was created by Toyota and now is being spread around the world with many other organizations also contributing to enrich those concepts and practices. In this chapter, three models were referred as including the right set of principles to be followed by those who want to also benefit from the wonders of CI. Nevertheless, CI is not easy to implement nor to master and a lot of hard work and perseverance is required to succeed with effectiveness. The main ingredients to implement CI are at least the following: The right mind set of the top management, a clear definition of the vision/ "True North", leadership system explicitly designed to create a CI culture, generation of natural teams, empowerment of natural teams, daily team meetings, visual management everywhere, supporting people with CI experience, always questioning the existing CI system, coaching skills spread around the organization, constant search for people engagement.

Unfortunately, very few organizations succeed in perfect align all the different required elements to create the right culture. In this line of thought my friend Frank Devine in one of his articles [14] also argue that most Corporate Leadership Programs are detached from, rather than integrated with, CI. These leadership programs are almost random collections of often unintegrated leadership skills rather than a system of leadership.

Acknowledgements

This work has been supported by FCT – Fundação para a Ciência e Tecnologia within the R&D Units Project Scope: UIDB/00319/2020".

English is not my native language so thank you to Frank Devine for bringing an Irishman's key eye for English grammar to my draft. And as he says.

Author details

Dinis Carvalho
University of Minho, Guimaraes, Portugal

*Address all correspondence to: dinis@dps.uminho.pt

References

[1] Sugimori Y, Kusunoki K, Cho F, Uchikawa S. Toyota production system and Kanban system materialization of just-in-time and respect-for-human system. International Journal of Production Research. 1977;**15**(6):553-564. DOI: 10.1080/00207547708943149

[2] Rother M. Toyota KATA. Europe: McGraw-Hill Education; 2010

[3] Krafcik JF. Triumph of the lean production system. Sloan Management Review. 1988;**30**(1):41-52. DOI: 10.1108/01443570911005992

[4] Womack J, Jones D, Roos D. The machine that changed the world. New York: Free Press; 1990

[5] Womack J, Jones D. Lean Thinking: Banish Waste and Create Wealth in Your Corporation. New York: Fee Press; 1996

[6] Plenert GJ. Discover Excellence: An Overview of the Shingo Model and its Guiding Principles. New York: CRC Press; 2017

[7] Shingo S. A Study of the Toyota Production System from an Industrial Engineering Viewpoint. New York: Productivity Press; 1989

[8] Shingo_Institute. Shingo Institute – Home of the Shingo Prize. 2019. Available from: https://shingo.org/ [Accessed: 20 January 2020]

[9] Liker J. Toyota Way: 14 Management Principles from the World's Greatest Manufacturer. New York: McGraw-Hill Education; 2004

[10] Toyota_Europe. The Toyota Way: our values and way of working. 2019. Available from: https://www.toyota-europe.com/world-of-toyota/this-is-toyota/the-toyota-way [Accessed: 20 January 2020]

[11] Yoshino I. Toyota Lean Leadership and Hoshin Kanri. 2019. Available from: https://www.slideshare.net/improven/toyota-lean-leadership-and-hoshin-kanri [Accessed: 21 May 2020]

[12] The_Shingo_Institute. MyEducator - The Shingo Institute. 2020. Available from: https://app.myeducator.com/reader/web/1705a/foreword/re0xx/ [Accessed: 18 August 2020]

[13] Liker J, Franz J. The Toyota Way to Continuous Improvement: Linking Strategy and Operational Excellence to Achieve Superior Performance. New York: McGraw-Hill Publishing; 2011

[14] Devine F. Demystifying Leadership Setting Leaders up for Success; 2015

[15] Pink D. Drive: The Surprising Truth about What Motivates Us. New York: Riverhead Books; 2009

[16] Heath C, Heath D. Switch: How to Change Things When Change Is Hard. New York; Broadway Books; 2010

[17] Covey SR. The 7 Habits of Highly Effective People. New York: Free Press; 2004

[18] Hamel MR, O'Connor M. Lean Math: Figuring to Improve. Dearborn, Michigan: SME; 2017

[19] Lloyd GC. Stuff the suggestions box. Total Quality Management. 1999;**10**(6):869-875. DOI: 10.1080/0954412997280

[20] Yasuda Y. 40 Years, 20 Million Ideas: The Toyota Suggestion System. Cambridge, MA, USA: Productivity Press; 1991

[21] Toyota_Motor_Corporation. TOYOTA MOTOR CORPORATION GLOBAL WEBSITE. 75 Years of

TOYOTA. 2012. Available from: http://www.toyota-global.com/company/history_of_toyota/75years/data/conditions/philosophy/toyotaway2001.html. [Accessed: 23 January 2020]

[22] Shingo S, Dillon AP. A Study of the Toyota Production System: From an Industrial Engineering Viewpoint. London: CRC Press; 1989

[23] Ahlstrom J. How to Succeed with Continuous Improvement: A Primer for Becoming the Best in the World. New York: McGraw-Hill Education; 2014

[24] Liker JK, Hoseus M. Human resource development in Toyota culture. International Journal of Human Resources Development and Management. 2010;**10**(1):34-50. DOI: 10.1504/ijhrdm.2010.029445

[25] Edgeman R, Barker S. Complex Management Systems and the Shingo Model: Foundations of Operational Excellence and Supporting Tools. New York: Productivity Press; 2019

[26] Sutherland J. Scrum: The Art of Doing Twice the Work in Half the Time. New York: Crown Business; 2014

[27] McNichols T, Hassinger R, Bapst GW. Quick and continuous improvement through kaizen blitz. Hospital Materiel Management Quarterly. 1999;**20**(4):1-7

[28] Bastos A, Sharman C. Strat to Action - O Método KAIZEN™ de levar a Estratégia à Prática. Porto, Portugal: Kaizen Institute; 2018

Chapter 4

A Service Management Metric with Origin in Plant Management

Robert G. Batson

Abstract

The discipline of industrial engineering (IE) was originated in the US by Frederick W. Taylor, who first applied what he termed scientific management to machine shop management in the 1880s. IE expanded world-wide with applications of work measurement to all manner of manufacturing, then to services. A century later, as the Japanese practice of Total Productive Maintenance (TPM) became known in the US, the associated equipment management metric Overall Equipment Effectiveness (OEE) became well-known as a production metric that could be applied to individual manufacturing machines, production lines, and the overall production system. In this chapter, illustrations of the calculation of OEE are provided, along with modification of the three OEE inputs (availability, performance efficiency, and quality rate) to create a new service management metric Overall Service Effectiveness (OSE). Definitions and measures of service quality are reviewed. The first published application of OSE was to a city public transportation system, and is reviewed as a prototype. Essentially, applications of OSE require the industrial engineer to define service-specific measures of availability, processing rate, and quality at the management's level of interest: work station, process, or system. The data collected to calculate OSE will also point toward actions that would improve OSE.

Keywords: service systems, service management, performance measurement, overall equipment effectiveness (OEE), overall service effectiveness (OSE)

1. Introduction

The discipline we now call industrial engineering (IE) originated in the US with the practices of Frederick W. Taylor at the Midvale Steel Company in the 1880s as he progressed from machinist, to time clerk, to machine shop foreman, ultimately becoming chief engineer upon receiving a mechanical engineering degree in 1883. His participation in the American Society of Mechanical Engineers (ASME) provided him with the opportunity to present his shop management practices which were referred to as "work measurement" when applied to a specific work task (manual labor such as shoveling; skilled labor such as lathe operation). Broader applications to groups of workers in a plant or service organization (educational organizations, government agencies, the ASME) became known world-wide as Scientific Management [1], especially after Taylor testified before the US Interstate Commerce Commission in 1911.

Henry L. Gantt was recruited to perform work measurement at Midvale Steel under Taylor's guidance, and as a consultant one speed and feed problems in metal cutting at Bethlehem Steel. Gantt modified one of Taylor's published practices (piece-rate system) to account for productivity factors outside the workers control. Gantt became an independent consultant and ultimately lectured on IE at four US universities. Another early practitioner of IE was Morris L. Cooke, whom Taylor funded to work on efficiency and effectiveness of the ASME, the Carnegie Foundation, and the municipal government of Philadelphia. Frank B. Gilbreth originated the practice of work measurement in the construction trades, though he never attended college. His approach came to be known as time and motion study, which he first applied to bricklaying (a trade he learned as an apprentice). He insisted on division of labor between the brick mason (skilled labor) and the unskilled workers who "set up" the mason with bricks and fresh mortar; the specific location of the bricks and mortar relative to the mason, and even the consistency of the mortar, could be planned to make the mason as productive as possible. Furthermore, with appropriate design of the motions the mason should use, he demonstrated that the mason could increase the number of bricks laid in a given time by a factor of three. At age 27, Gilbreth founded (1895) a highly successful construction firm wherein all work was designed using time and motion study, but gave it up at age 44 to become a full-time management consultant. Frank's wife, Lillian M. Gilbreth, was a PhD psychologist who assisted Frank in the preparation of six books between 1908 and 1917 to disseminate what he had learned about the broad topic of performance measurement, starting with the worker and broadening to the work processes and the overall work system.

As Japan began to recover from destruction of its industrial base during WWII, and to transition from essentially an agrarian society to an economic powerhouse, their industrial/production engineers originated many practices now considered part of modern industrial engineering. Starting in the 1970s and intensifying in the 1980s, there was significant debate in the US and other advanced economies in the West concerning what was enabling the Japanese to capture larger and larger market share in technological products such as automobiles, televisions, and copy machines. There was a US IE professor, Richard J. Schonberger, who spend a significant amount of time in Japan and authored several books [2–4] detailing his interviews and observations from visiting top-performing Japanese manufacturing firms. In *Japanese Manufacturing Techniques* [2], he revealed the following nine "hidden lessons in simplicity":

1. Fewer suppliers

2. Reduced part counts

3. Focused factories (focus on a narrow line of products)

4. Scheduling to a rate, instead of scheduling by lots

5. Fewer racks on the plant floor

6. More frequent deliveries (in-plant moves, as well as deliveries from suppliers)

7. Smaller plants

8. Shorter distances, less reporting, less inspectors, less buffer stock

9. Fewer job classifications.

In *World Class Manufacturing* [3], Schonberger claimed production management in the US had become overly focused on "managing by the numbers" by which he meant measuring plant performance at too high a level (revenue, fixed and variable costs, profit) to really uncover hidden efficiency and quality effects; whereas in contrast, WCM mandates simplification and direct action—do it, judge it, measure it, diagnose it, fix it, manage it, on the factory floor. He provided a variety of examples with diagrams/photographs in [4], *World Class Manufacturing Casebook*. In [3] Schonberger observed that in 1980, the first US WCM thrusts followed two parallel paths: a quality path with a goal of zero defects known in Japan at Total Quality Control (TQC) and in the US as Total Quality Management (TQM); a just-in-time (JIT) productivity path, as a means of coping with high-variety, small lot, short lead-time production. JIT aims to have every operation make the needed items at the right time in the right quantities, at low cost. JIT pursues a goal of one-piece flow, small-lot production, with minimal inventory throughout the system. A third WCM practice which supports both TQC and JIT is known as Total Productive Maintenance (TPM) and will be described below when we introduce the metric Overall Equipment Effectiveness (OEE). The three practices are synergistic—for instance, JIT will fail if incoming part quality is not (nearly) perfect and processing equipment does not have OEE near 100%.

Most US industrial engineers first learned details of TPM through the 1988 English version of Nakajima's *TPM: Introduction to Total Productive Maintenance* [5]. According to Nakajima "TPM is an innovative approach to maintenance that optimizes equipment effectiveness, eliminates breakdowns, and promotes autonomous operator maintenance through day-to-day activities involving the total workforce." Seven years later, two North American practitioners Charles Robinson and Andrew Ginder produced the well-known *Implementing TPM: The North American Experience* [6] and therein defined TPM to be "a plant improvement methodology which enables continuous and rapid improvement of the manufacturing process the use of employee involvement, employee empowerment, and closed-loop measurement of results." In the decade following Robinson and Ginder [6], publications on "TPM Practices and Cases" [7] and "Lean TPM" [8] appeared in the US and Great Britain. In the recent past, Ortiz contributed *The TPM Playbook: A Step-by-Step Guideline for the Lean Practitioner* [9] and Peng, in response with the digital revolution in production, published *Equipment Maintenance in the Post-Maintenance Era: A New Alternative to Total Productive Maintenance (TPM)* [10]. All of these references include a section or entire chapter explaining the measurement of Overall Equipment Effectiveness; furthermore, all suggest an engineer or improvement team could focus on OEE for a single machine (workstation), OEE for a production line or process, and OEE for the overall plant. Comparisons across shifts, days, months, etc. would detect improving or deteriorating performance. Furthermore, comparisons across similar machines, processes, or plants could be very useful to department, process, or plant managers.

OEE for a given machine, line, or plant is the product of availability, performance efficiency (processing rate ratioed with the design "ideal"), and quality rate (proportion of good products produced—the yield). Because each of these inputs is measured as a percentage, the closer OEE is to 100%, the better; world-class OEE is considered 85% or higher (some authors say 90% or higher). OEE is formally calculated using the following expressions, each expressed as a percentage:

$$\text{Availability} = (\text{loading time} - \text{downtime}) / (\text{loading time}) \tag{1}$$

$$\text{Performance Efficiency} = (\text{theoretical cycle time} \times \text{processed amount}) / (\text{operating time}) \tag{2}$$

$$\text{Quality Rate} = (\text{processed amount} - \text{defect amount}) / \text{processed amount} \tag{3}$$

$$\text{OEE} = \text{Availability} \times \text{Performance Efficiency} \times \text{Quality Rate}. \tag{4}$$

The calculation of each of these quantities is illustrated by example in the references by Nakajima [5], and Robinson and Ginder [6]. The example in Nakajima further illustrates why the three input quantities to OEE are seldom calculated to be 100%. Essentially, the use of OEE in TPM uncovers the "Six Big Losses" which become the focus of improvement efforts by an individual engineer, or a team. The Six Big Losses are grouped as follows:

Losses that determine equipment availability

1. Equipment failure losses (requiring corrective maintenance)

2. Set-up and adjustment losses

Losses that determine performance efficiency

3. Idling and minor stoppages (e.g., clearing a jammed workpiece, or stopping for a visitor)

4. Reduced speed (e.g., running slower to avoid overheating, or avoid early job completion)

Losses that determine rate of quality products

5. Defects and rework

6. Reduced yield due to start-up losses (either due to nature of process, or company policy)

Some examples of how to improve OEE above status quo, based on the six big losses as numbered above, would be:

1. Study equipment failure and repair records. Use the Pareto Principle to identify which machines are causing downtime, in rank order: then, for the least available machine, identify the machine elements that are the causes of downtime, in rank order. Focus improvement efforts on the most problematic machines, and once identified, the most problematic machine elements.

2. To reduce set-up time, there is a Japanese practice originally known as Single Minute Change of Die (SMED)—meaning work to reduce change-over times to less than 10 minutes (single digits) with a goal of single minute change-overs; in the US, this practice is called Quick Changeover Technology and, for example, has been observed in casting machines in pipe shops, and welding machines for tubular steel products. This practice is essentially a specialized time and motion study, again carried out by a single engineer or improvement team, and the saving from avoided set-up losses can be substantial.

3. Periods of idling and instances of minor stoppages should be recorded (total time lost, situation and/or causes).

4. Reduced speed losses—there may be good reasons for running equipment at less than ideal processing rate, such as to avoid overstressing the equipment or safety concerns for the operator or other workers in the vicinity. In instances where work crews intentionally slow down, management needs to re-plan schedules so every worker can get in a full shift, having come to work intending to be paid for a full shift.

5. Defects and rework must be recorded and carefully examined to determine root causes, and then immediate corrective actions taken (standards modified or adjusted) to hopefully prevent the same problems in the future. Follow-up is critical by the engineer, manager, or team to verify the action installed is working and has become the standard.

6. Start-up losses may be unavoidable with the materials, machine, and set-up required; or, they may indicate a company policy that is outdated (current machine *can now* produce acceptable quality with first unit produced, or *could* with appropriate attention from engineering, maintenance, and production).

For more details on data collection and the calculation/application of OEE, see three references focused specifically on OEE: Muchiri and Pintelon's 2008 article "Literature Review and Practical Application Discussion" [11]; Hansen's *Overall Equipment Effectiveness* book [12]; and *The OEE Primer: Understanding Overall Equipment Effectiveness, Reliability, and Maintainability* [13] by Stamatis. As with the world-wide spread of Scientific Management, the use of OEE by industrial engineers has spread to practitioners in Europe (e.g., see an application in France to large-scale production of ductile iron pipe [14], and in Asia, for example, applications to sugar mills in India [15, 16]).

2. Service management

A simple definition of service is "work performed for someone else." In other words, services are all those economic activities in which the primary output is neither a good nor a construct, so services are for the most part intangible. Of course, services may occur *internal* to a company, educational institution, medical facility, or government agency, or in service encounters between individuals, or exchanges between larger-scale entities—two companies, or perhaps a university and a government agency. In general, services cannot be inventoried. When considering services in context of the overall economy, a big distinction is that product-oriented sectors of an economy always produce a tangible product; a service may or may not terminate in a tangible product. Also, services are rendered on demand—either instant demand or scheduled demand—often with the customer present and involved in the service. So, the reliability characteristic "ready on demand" is critical to high quality service. Once service begins, uninterrupted service may be another customer expectation. Variability of service in response to specific customer requests may be intentional—information gets transformed into customized action in an attempt to satisfy the requests. It may also be unclear which party "owns" the service. In contrast, a manufactured product is tangible, produced and consumed at different locations at different times, expected to be of consistent quality, to be inventoried with like items, and to have clear ownership as it changes hands.

The Service Sector of the US economy in 2019 accounted for 79% of US employment. The percentage of service workers in other advanced economies are approximately 75% in Great Britain, 65% in France, and 60% in Germany and Japan. Service is considered a tertiary sector following the Primary Sector "Extractive" industries (Fishing, Agriculture, Mining, Oil and Gas, etc.) and the Secondary Sector "Transformative" industries (Manufacturing and Construction). The service sector is highly diverse; see the following groupings used by the US Dept. of Labor:

- Trade, Transportation, and Utilities
    - o Wholesale Trade (NAICS 42)
    - o Retail Trade (NAICS 44-45)
    - o Transportation and Warehousing (NAICS 48-49)
    - o Utilities (NAICS 22)
- Information (NAICS 51)
- Financial Activities
    - o Finance and Insurance (NAICS 52)
    - o Real Estate and Rental and Leasing (NAICS 53)
- Professional and Business Services
    - o Professional Scientific and Technical Services (NAICS 54)
    - o Management of Companies and Enterprises (NAICS 55)
    - o Administrative and Support; Waste Management and Remediation (NAICS 56)
- Education and Health Services
    - o Educational Services (NAICS 61)
    - o Health Care and Social Assistance (NAICS 62)
- Leisure and Hospitality
    - o Arts, Entertainment, and Recreation (NAICS 71)
    - o Accommodations and Food Services (NAICS 72)
- Other Services (except Public Administration) (NAICS 81)
- Government

Characteristics of services are:

- The product is intangible, for the most part
- Usually performed in real time, with the customer present and often participating
- Seldom inventoried, so must be delivered on the customer's schedule
- Something of value is provided (like manufacturing) but in a more immediate, personalized manner.

Quality of a product (service, good, or software) has been defined by Juran [17] as "fitness for use, in the intended environment" and by Deming [18] as "meets or exceeds customer expectations." Therefore, a broad definition of service quality might be "fitness for use as determined by those features of the service that the customer considers to be beneficial." As explained in Jain and Gupta [19], "services require a distinct framework for quality explication and measurement" and "involves evaluation of the outcome (i.e., what the customer actually receives from the service) and the process of service act (i.e., the manner in which the service is delivered)." Another line of research at that time concerned measurement of service expectations, pre- and post-consumption [20]. Parasuraman et al. [21] created the SERVQUAL scales and questionnaire organized around these five measures of service quality:

- Reliability—the ability to perform the desired service dependably and accurately
- Assurance—the knowledge and courtesy of employees and their ability to convey trust and confidence
- Tangibles—the appearance of physical facilities, equipment, personnel, and communication materials
- Empathy—the provision of caring, individualized attention to the customer
- Responsiveness—the willingness to help customers and provide prompt service.

In a later publication [22], these same three authors presented a list of nine dimensions classifying how customers perceive service quality:

- Tangibles—physical appearance
- Reliability—performed as promised, consistently
- Responsiveness
- Competence
- Credibility
- Security/Safety
- Access—easy to do business with
- Communication—keeping customer informed
- Understanding customer needs.

Pitt et al. [23] concluded that SERVQUAL is an appropriate instrument for researchers seeking a measure of information system service quality.

We have collected together six widely-known customer reactions to service quality:

- Poor or inattentive service costs companies about 10% of volume annually (until corrected)
- 96% of unhappy customers never complain, but 90% never return. Each unhappy customer tells at least nine others.

- Each happy customer tells five others, who may become future customers.
- The best opportunity to increase sales and market share is through your present customer base
- Customer perception of quality of service depends heavily on employee's job satisfaction (dis-satisfaction)
- Service personnel are *critically dependent on systems* (often computer-based) to deliver quality to the customer.

Let us consider several examples of the last point, which is very important for the industrial engineer:

- A college instructor depends on:
  - o The student registration system to provide accurate class rolls and a means to report final grades to the registrar
  - o The classroom assignment system to provide a lecture hall to match the enrollment
  - o The textbook ordering system to order, receive, and distribute the correct class materials in the correct quantities, on time
  - o The classroom audio-visual system and computer software/internet access provided in the lecture hall.
- A medical doctor seeing patients in his/her clinic depends on:
  - o The patient appointment scheduling system
  - o The measurement of patient vital signs by nurses as the visit begins, with computer access
  - o Blood testing machines and/or radiological scans done prior to the visit, with computer access
  - o Computer access to records of previous ailments and treatments, surgeries, vaccinations, etc.
  - o Equipment he/she may use during the patient encounter.
- A service representative at a cable television/internet provider depends on:
  - o An information system showing the customer's current service details, including start-up and service end dates
  - o An information system showing additional or alternative services available to the customer based on location, with costs and time frame for change in service
  - o A billing system should balances in accounts, due dates, penalties for late payments, etc.
- A bank teller for customers who walk in the branch and queue for service, depends on:

- An information system showing a customer's accounts, safe deposit boxes, progress on any money electronically moved from or to customer's accounts
- An information system tracking cash transactions (deposits, dispersals, exchanges, etc.) completed by the teller and what should be the status of their cash drawer
- In a large bank, a formal schedule of teller work assignments and for each, their schedule of breaks and lunch, and an out-of-office schedule for the current day and perhaps for the weeks or months ahead.

3. Overall service effectiveness

Overall Service Effectiveness (OSE) was first described in Berhan [24], and is the focus of the remainder of this chapter. The OSE metric for services extends the OEE production metric developed along with TPM as described earlier in the chapter. For the reader's convenience, we shall demonstrate how OSE is a simple rewrite of the formulas for OEE and its three input components in a manner that fits service transactions an industrial engineer might be challenged to design or improve, using OSE as a guide.

In the equations below, the term "units" could be units of a manufactured good or quantities of a service completed. Examples of the latter might be: queries to an information systems; patients seen by a doctor or dentist; customer transactions at a bank—in person or electronic; riders transported by a bus or aircraft, or by the bus line or airline. Note these are all situations which an industrial engineer might encounter, and traditional IE tools such as queuing theory or system simulation might be in use. Agreeing with the OSE equations of Berhan [24], we shall use:

$$\text{Availability} = (\text{Total Time Available} - \text{Downtime}) / (\text{Total Time Available}) \tag{5}$$

$$\text{Performance} = (\text{Number of Units Produced}) / (\text{Ideal Number of Units Produced}) \tag{6}$$

$$\text{Quality} = (\text{Number of Units Produced} - \text{Number of Defects}) / (\text{Number of Units Produced}) \tag{7}$$

$$\text{OSE} = \text{Availability} \times \text{Performance} \times \text{Quality}. \tag{8}$$

An example for an urban transportation system described in Berhan [24] adapted the equations for the three inputs above to the specifics of the service operation as follows:

$$\text{Availability} = (\text{Total Transport Time Available} - \text{Lost Transport Time}) / (\text{Total Time Available}) \tag{9}$$

$$\text{Performance} = (\text{Number of Passengers Transported}) / (\text{Target Number of Passengers Transportable}) \tag{10}$$

$$\text{Quality} = (\text{Number of Passengers Transported} - \text{Number of Dissatisfied Passengers}) / (\text{Number of Passengers Transported}) \tag{11}$$

Note that the bus service, like many encountered in modern society, is a "knowledge embedded service" [25] which are services which embed the customer value in a system that provides the service, so human-machine system reliability is a key component of the availability input to OSE for such services. Here, the driver's knowledge of the route and how to operate the bus matters as much as the bus reliability.

Using real data from the public transport (bus) system in Addis Ababa, Ethiopia: Berhan first computed planned downtime (lunch breaks and shift changes), downtime and speed losses, performance efficiency losses, and finally quality and yield losses; Berhan then computed:

$$\text{Availability} = 78.85\% \quad (12)$$

$$\text{Performance} = 74.89\% \quad (13)$$

$$\text{Quality} = 70.83\% \quad (14)$$

which yielded a system - wide effectiveness measure of

$$\text{OSE} = 78.85\% \times 74.89\% \times 70.83\% = 41.83\% \quad (15)$$

showing this service system needs significant improvement in order to be rated "world-class".

Just like in manufacturing, the OSE could also have been calculated for each bus individually, or for groups of busses that act together to cover a given route or sector within the city. Hence, OSE would be useful for service performance improvement at the bus, route, or (as demonstrated) system level.

4. Conclusions

This chapter provided background on the application of work measurement to services, starting with Taylor and his associates, and tracing the evolution from the plant performance metric Overall Equipment Effectiveness to an innovative service performance metric Overall Service Effectiveness (OSE). As illustrated in the analysis of an existing city bus system, the details used to compute the OSE inputs (availability, performance efficiency, quality rate) point toward actions that would improve OSE toward 100%. When designing a new service system (e.g., bus line, bank layout, fast food restaurant) the OSE metric can be used along with other industrial engineering tools (e.g., classic queuing formulas, systems simulation, engineering economy) to arrive at the most cost-effective layout, equipment/ software, and staffing to handle forecast service demands.

Author details

Robert G. Batson
The University of Alabama, Tuscaloosa, Alabama, USA

*Address all correspondence to: rbatson@eng.ua.edu

References

[1] Taylor FW. Principles of Scientific Management. New York: Harper & Brothers; 1911

[2] Schonberger RJ. Japanese Manufacturing Techniques: Nine Hidden Lessons in Simplicity. New York: The Free Press; 1982

[3] Schonberger RJ. World Class Manufacturing. New York: The Free Press; 1986

[4] Schonberger RJ. World Class Manufacturing Casebook. New York: The Free Press; 1987

[5] Nakajima S. TPM: Introduction to Total Productive Maintenance. Cambridge, MA: Productivity Press; 1988

[6] Robinson CJ, Ginder AP. Implementing TPM: The North American Experience. Cambridge, MA: Productivity Press; 1995

[7] Bernstein RF, editor. TPM Collected Practices and Cases. Cambridge, MA: Productivity Press; 2005

[8] Rich N, McCarthey D. Lean TPM: A Blueprint for Change. 2nd ed. Waltham, MA: Butterworth-Heinemann; 2015

[9] Ortiz CA. The TPM Playbook: A Step-by-Step Guideline for the Lean Practitioner. New York: Productivity Press; 2016

[10] Peng K. Equipment Maintenance in the Post-Maintenance Era: A New Alternative to Total Productive Maintenance (TPM). Boca Raton, FL: CRC Press; 2018

[11] Muchiri P, Pentelon L. Performance measurement suing overall equipment effectiveness (OEE): Literature review and practical application discussion. International Journal of Production Research. 2008;**46**(13):3517-3535

[12] Hansen RC. Overall Equipment Effectiveness. New York: Industrial Press, Inc.; 2011

[13] Stamatis DH. The OEE Primer. Boca Raton, FL: CRC Press; 2010

[14] Charaf K, Ding H. Is overall equipment effectiveness universally applicable? The case of Saint-Gobain. International Journal of Economics and Finance. 2015;**7**(2):241-252

[15] Singh K, Singh S. Eliminating obstacles in TPM implementation in sugar mill industry. International Journal of Advance Research and Innovation. 2018;**6**(4):332-334

[16] Singh S, Szzingh K, Mahajan V, Singh G. Justification of overall equipment effectiveness (OEE) in Indian sugar mill industry for attaining core excellence. International Journal of Advance Research and Innovation. 2020;**8**(1):34-36

[17] Juran JM. Juran on Leadership for Quality. New York: The Free Press; 1989

[18] Deming WE. Out of the Crisis. Cambridge, MA: The MIT Press; 1982

[19] Jain SK, Gupta G. Measuring service quality: SERVQUAL vs. SERVPERF scales. Vikalpa. 2004;**29**(2):25-38

[20] Polonsky M, Higgs B, Hollick M. Measuring expectations: Pre and post consumption: Does it matter? Journal of Retailing and Consumer Services. 2005;**12**(1):49-64

[21] Parasuraman A, Zeithaml V, Berry LL. SERVQUAL: A multiple-item scale for measuring customer perceptions of service quality. Journal of Retailing. 1985;**62**(1):12-40

[22] Zeithaml V, Parasuraman A, Berry LL. Delivering Service Quality: Balancing Customer Perceptions and

Expectations. New York: The Free Press; 1990

[23] Pitt LF, Watson RT, Bruce Kavan C. Service quality: A measure of information system effectiveness. MIS Quarterly. 1995;**19**(2):173-187

[24] Berhan E. Overall service effectiveness on urban public transport system in the City of Addis Ababa. British Journal of Applied Science and Technology. 2016;**12**(5):1-9

[25] Krause DR, Scannell TV. Supplier development practices: Product- and service-based industry comparisons. The Journal of Supply Chain Management. 2002;**38**(2):13-21

Chapter 5

Industrial Safety Management Using Innovative and Proactive Strategies

Siyuan Song and Ibukun Awolusi

Abstract

Safety is considered a top priority due to its significance in safeguarding human lives and properties, especially in high-risk industrial sectors such as aviation, oil and gas, construction, transportation, steel manufacturing, and mining industries. These industries are plagued by workplace injuries, illnesses, and fatalities because of the dangerous work environments. As such, it is very vital to integrate safety into every work process in any industrial environment just like quality is built into products and services. It is important to establish and execute an effective safety management system to prevent the risks of irreversible accidents. This chapter begins with a background to safety management in industrial engineering and a discussion of the various issues of industrial safety management. It follows with an extensive description of existing and commonly used safety performance measurement methods. Several case studies are used to explain the methods and explore the important application areas relevant to most industrial sectors. The techniques and tools for safety data collection, analysis, and sharing are introduced together with their applications for safety management. The last section explains how emerging technologies can be implemented in most industrial sectors to enhance safety management.

Keywords: emerging technologies, hazard identification, safety management

1. Introduction

Industrial work environments are often characterized by dynamic resources including interactions between mobile equipment and pedestrian workers. The hazardous work environment characteristic of industrial facilities is evident in the high rates of workplace injuries and fatalities experienced regularly. These high-risk industries include construction, steel manufacturing, oil and gas, aviation, agriculture, forestry, fishing, and hunting, etc.

For instance, the construction industry remains one of the most hazardous and unsafe industries with fatality and incidence rates considerably higher than the all-industry average in many countries [1–4]. Incident statistics indicate that construction workers have consistently incurred more fatal injuries than in other industries. Despite the efforts to improve safety performance, the construction sector continues to account for disproportionate injury rates accounting for the most on-the-job fatal injuries. In the United States, construction remains the most hazardous industry in terms of the aggregate number of fatalities [1]. Thus, innovative intervention strategies are being continuously explored by researchers and practitioners to enhance management controls as well as modify human behavior and work environment to improve construction safety.

Steel manufacturing is one of the most hazardous industries because of its complex socio-technical system. The steel manufacturing process involves the use of high technology and physical labor, making safety management a complicated task [5]. Members of the U.S. steel manufacturing industry continue to experience a significant number of injuries, illnesses, and fatalities [6]. The combination of intricate technology and physical labor creates a complicated challenge for safety managers in steel manufacturing [5].

The fundamental goal of measuring safety performance is to create and implement intervention strategies for potential avoidance of future accidents. Recognizing signals before an accident occurs offers the potential for improving safety; many organizations have sought to develop programs to identify and benefit from alerts, signals, and prior indicators [7]. Traditional measures of safety performance rely on some form of accident or injury data [8], with actions being taken in response to adverse trends in injuries [9]. Many organizations rely heavily on failure data to monitor performance. The consequence of this approach is that improvements or changes are only determined after something has gone wrong [10]. In most cases, the difference between whether a system failure results in a minor or catastrophic outcome is purely a matter of chance.

Effective management of major hazards requires a proactive approach to risk management, so information to confirm that critical systems are operating as intended is essential [11]. Transitioning the emphasis in favor of leading indicators to confirm that risk controls continue to operate is an important step forward in the management of major hazard risks [10]. Accurate safety performance measurement facilitates the evaluation of ongoing safety management and the motivation of project participants to improve safety [12].

The ability to collect, analyze and disseminate safety information using a large amount of useful data from leading indicators can allow for hazardous events and conditions to be efficiently mitigated and controlled before a lagging indicator occurs [11]. In this chapter, a background to safety management in industrial engineering is presented followed by a discussion of the various issues of industrial safety management. The existing and commonly used safety performance measurement methods are extensively described. Several case studies are used to explain the methods and explore the important application areas relevant to most industrial sectors. The techniques and tools for safety data collection, analysis, and sharing are introduced together with their applications for safety management while the use of emerging technologies for enhancing safety management in most industries is discussed in the last section.

2. Safety culture

The safety culture of an organization refers to the product of individual and group values, attitudes, perceptions, competencies, and patterns of behavior that determine the commitment to and the style and proficiency of an organization's safety and health management [13]. Safety culture has been defined in a variety of ways and there is no standard definition of safety culture. This is mainly because a culture of safety has diverse meanings in different industries and people may have

Reference	Definition of safety culture
[8]	Safety culture is thought to influence employees' attitudes and behavior in relation to an organization's ongoing health and safety performance [14]
[15]	The Safety Culture is made up of a collection of individual cultures and other subcultures within the environmental constraints and promotions of the organization [15]
[16]	Safety culture is defined as a set of prevailing indicators, beliefs, and values that the organization owns in safety [16]
[17]	Safety culture is a sub-facet of organizational culture, which is thought to affect members' attitudes and behavior in relation to an organization's ongoing health and safety performance [17]
[18]	Safety culture forms a subset of organizational culture relating specifically to the values and beliefs concerning health and safety within an organization [18]
[19]	Safety culture refers to shared attitudes, values, beliefs, and practices concerning safety and the necessity for effective controls [19]
[20]	Safety culture is defined as: those aspects of the organizational culture which will impact on attitudes and behavior related to increasing or decreasing risk [20]
[21]	Safety culture is shaped by people working together in organizational structures and social relationships in the workplace. The key attributes of organizational culture are defined as organizational communication, senior management commitment and organizational learning [21]

Table 1.
Selected safety culture definitions.

various understandings in different situations. Selected examples of safety culture definitions are organized and shown in **Table 1**.

Strong safety culture has a significant impact on improving safety performance, reducing incidents, conducting a successful near-miss, and incident reporting in an organization. The growing importance of safety culture to the industry is evidenced by reports, guidelines, publications, workshops, and conferences. As an industry-led initiative, the Center for Offshore Safety (COS) defined six specific elements characteristic of a successful offshore safety culture, including leadership, respect and trust, environment for raising concerns, open communication, personal accountability, and inquiring attitude [22]. According to the European Union Agency for Railways (ERA), "Safety culture refers to the interaction between the requirements of the Safety Management System (SMS), how people make sense of them, based on their attitudes, values, and beliefs, and what they actually do, as seen in decisions and behaviors" [23]. A safety culture model (**Figure 1**) was developed by ERA to assess safety culture and identify improvable areas [23]. The model is made up of three building blocks: cultural enablers, behavior patterns, and railway safety fundamentals [23].

- Cultural enablers: those levers through which an organizational culture develops;

- Behavior patterns: those shared ways of thinking and acting which convey the organizational culture;

- Railway safety fundamentals: those core principles which must be reflected by behavior patterns to achieve sustainable safety performance and organizational excellence.

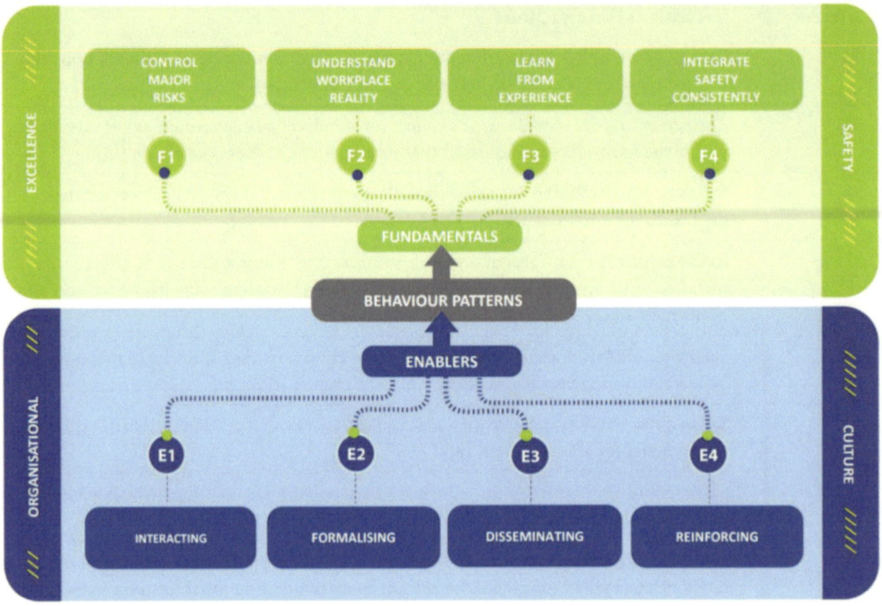

Figure 1.
European Railway Safety Culture Model 2.0: Components.

Within a positive safety culture, the organization's formal management systems and leaders' informal management practices encourage, recognize, and reinforce safe behaviors, and create an environment where employees feel responsible for their safety and the safety of their peers [14]. The largest indicator of a management's commitment to safety is the investments made for safety including discretionary safety funding [24]. Previous research investigated the correlation between safety discretionary funding of construction companies and their corresponding safety record [25]. Results suggest that increasing the amount of discretionary safety funding in a construction company can improve their incident record. Furthermore, companies that invest in safety programs, training, and employee incentives can improve their safety record [25]. Finally, results from a construction safety study found that organizational commitment throughout all levels (top management, site level, to the individual level) is the key to promoting improved safety performance [26].

3. Safety performance measurement

The primary goal of measuring safety performance in a work environment is to intervene in an attempt to mitigate unsafe behaviors and conditions that can lead to accidents. Various measures of safety performance have been used for decades and they have served a useful purpose [27]. Generally, performance measurements can either be reactive or active monitoring [10]. While reactive monitoring means identifying and reporting on incidents and learning from mistakes, active monitoring provides feedback on performance before an accident or incident occurs [28].

In the US, safety performance has traditionally been measured by metrics such as the Occupational Safety and Health Administration (OSHA) recordable injury rate (RIR); days away, restricted work, or transfer (DART) injury rate; or the experience modification rating (EMR) on workers' compensation [27]. Past safety performance has been largely measured and driven by lagging indicators (including injuries, illnesses, and fatalities), but improvements and enhancements of safety performance can be experienced through implementing safety leading indicators to measure worker safety performance [11]. Although lagging indicators will continue to be used, they have serious limitations when it comes to the prediction of the current and future safety performance of a project or work environment. This makes the need for leading indicators of safety performance very crucial [27].

The term "indicators" is used to mean observable measures that provide insights into a concept that is difficult to measure directly; a safety performance indicator is a means for measuring the changes over time at the level of safety as the result of actions taken [29]. An indicator is a measurable and operational variable that can be used to describe the condition of a broader phenomenon or aspect of reality. An indicator can be considered any measure (quantitative or qualitative) that seeks to produce information on an issue of interest [30]. Safety indicators can play a key role in providing information on organizational performance, motivating people to work on safety, and increasing the organizational potential for safety [31].

The major distinction between leading and lagging indicators lies in the type of response that is elicited by them when the measures indicate that performance is not as desired. While in leading indicators, the response is proactive in nature with the intent of making changes in the safety process to avoid injuries, with lagging indicators, the response is reactive as a response is made after injuries have already occurred and the response is initiated to try to prevent the occurrence of further injuries [27]. Hence, the two categories of safety metrics are: (1) lagging indicators (i.e. metrics linked to the outcome of an injury or accident); and (2) leading indicators (i.e. metrics or measurements linked to preventive actions).

3.1 Lagging indicators

Lagging indicators are related to reactive monitoring which involves identifying and reporting on incidents to check that controls in place are adequate, to identify weaknesses or gaps in control systems, and to learn from mistakes [10]. They show when the desired safety outcome has failed, or when it has not been achieved [32]. Since lagging indicators might prompt response after an injury or a series of injuries have occurred, it should be evident that lagging indicators of safety performance are based on past safety performance results [27]. Lagging indicators do not provide further insights on the existing safety conditions once an accident has occurred because they do not give room for informed decision making based on continuous data collection and analysis [33]. The most commonly used lagging indicators are accident rate, lost workday injuries, medical case injuries, and experience modification rate (EMR), among others.

3.2 Leading indicators

Leading indicators are measurements of processes, activities, and conditions that define performance and can predict future results [9]. A leading indicator is the result of periodic measurements of specific safety performance. Leading indicators provide opportunities for safety managers to identify areas of safety performance that need improvement before injuries or fatalities occur [34]. Leading indicators measure the building blocks of the safety culture of a project or company. When one or more of these measures suggest that some aspect of the safety process is weak or weakening, interventions can be implemented to improve the safety process and, thereby positively impact the safety process before any negative occurrences (injuries) are sustained [9]. The common leading indicators used in industrial sectors are near miss reporting, project management team safety process involvement, worker observation process, job site audits, housekeeping program, stop work authority, safety orientation and training, etc. Leading indicators consist of both passive as well as active measures. Passive measures are those which can be predictive over an extended period while active measures are those which can initiate corrective steps in a short period. These two measures of leading indicators are further described as follows.

3.2.1 Passive leading indicators

Passive leading indicators are those that provide an indication of the probable safety performance to be realized within a firm or on a project. While they may be somewhat predictive on a macro scale, they are less effective as being predictive on a short-term basis. This implies that the process being monitored by passive leading indicators cannot generally be altered in a short period of time [27]. Measures of passive indicators are usually binary in that the organization implements them or does not [35]. The most reliable information that passive indicators provide when properly analyzed and applied is a simple qualitative measure of the knowledge or skills base of personnel which is useful in implementing a comprehensive safety management system [27].

3.2.2 Active leading indicators

Active leading indicators are those which are more subject to change in a short period. Active leading indicators can either be quantitative, but the measures can also be qualitative. Quantitative measures may be preferred as they are more objective and may result in more consistent interpretation. Nonetheless, when no other means are available, qualitative measures should not be avoided [27]. The leading indicators of safety performance essentially disclose what aspects of the safety program are going well and, if there are any weaknesses, these will be identified, and implementation of change can be initiated. Active indicators are generally continuous in that they occur at a frequency or are measures of quality of implementation [35]. Active leading indicators represent both a qualitative and quantitative measure of the actual implementation of the processes within a comprehensive safety management system [27].

4. Safety management through hazard identification

In high-risk industrial sectors, workers are constantly exposed to various types of occupational hazards due to the nature of their work and condition of the work environment. The first step in accident prevention is the identification of hazards. The improvement of safety performance requires the implementation of proactive worker hazard identification and prevention programs. In many industrial sectors, the safety performance of workers is predominantly measured based of their ability to proactively identify and respond to hazards in the work environment [36]. Hazards are related to the improper release of energy. Accidents result from the interaction of energy, equipment or materials, and one or more people, and the potential hazards associated with such interaction can be identified based on the energy sources recognition.

Being oblivious of the presence and magnitude of an energy source often results in an accident. As a result, it is important to identify highly innovative and effective hazard recognition strategies such as implementing techniques to avoid future accidents [36]. Because hazards can be caused by different energy sources, the awareness of all the energy sources is key to identifying potential hazards and creating a safe environment. The risk associated with hazardous conditions or situations in a work environment can only be analyzes for accident prevention if the related hazards can be correctly recognized or identified.

In construction, for instance, workers are constantly exposed to hazards that are difficult to measure due to the nature of the work environment and the way construction tasks are performed [28, 37]. Statistics indicate that fatalities and incidents rates in the construction industry remain significantly higher than the all-industry average in many countries. This makes the construction industry one of the most hazardous and accident-prone industries and the majority of the accidents experienced occur due to the inability of the workforce to predict, identify, and respond to hazards at the workplace [38]. The dynamic nature of construction work and task unpredictability on projects make hazard recognition difficult [39]. The energy required to accomplish work tasks on projects if released inappropriately may cause loss-of-control which can get construction workers injured [40]. The probability of accidents will increase when hazards are not identified and assessed on a typical project.

In the steel manufacturing industry, employees continuously work in highly hazardous work environments characterized by limited visibility, hazardous proximity situations between heavy equipment and pedestrian workers, and the dynamic nature of manufacturing tasks. The working conditions typical of steel manufacturing environments include increased amounts of repetitive work tasks [41], elevated temperature [42], noisy surroundings [43] and an overall rugged work environment [44]. These conditions tend to cultivate conditional and behavioral hazards that increase the probability of employees experiencing an incident in the form of an injury, illness, or fatality. The choice and the implementation of specific measures for preventing workplace injury and illness in the iron and steel industry depend on the recognition of the principal hazards and the anticipated injuries and diseases, ill health, and incidents [11]. Although hazard identification provides a useful method for mitigating hazards, the impact of specific hazards categories on injuries, illnesses, and fatalities has not been quantified [44]. To proactively identify hazardous situations and conditions, details from safety incident data can be analyzed to identify predictor variables of future incidents in steel manufacturing environments.

4.1 Energy source recognition

An injury occurs whenever energy is released from one or more of these sources and transferred to the human body. For instance, a suspended load is a source of gravity and motion because it has the potential to fall and swing. If a worker is struck by this suspended load, motion energy will be transferred from the load to the worker and absorbed by the worker's body, causing an injury. Other examples of energy sources in industrial work environments include radiation from welding, hot and cold objects and environments, compressed gas cylinders, hazardous substances, moving and noisy equipment and vehicles, and objects and bodies at height [45].

Certain hazard sources and activities may be associated with multiple hazards. For example, an electric cable on the floor may be associated with a trip hazard and an electrical hazard. Industrial work environments may contain hidden or dormant hazards that are not expected or perceived as imposing any imminent danger. Such hazards often remain in work environments as latent or stored energy for extended periods without causing any harm. However, the unexpected release or trigger of these latent sources of stored energy can result in dramatic injury and illnesses.

Because of the importance of hazard recognition, employers adopt several methods to improve hazard recognition levels. One of these methods is the retrospective hazard recognition method which is based on deducing or extrapolating knowledge gained and lessons learned from past safety incidents (i.e. accidents data) to new situations and projects [46, 47]. Despite these significant advancements, there is still a dearth of research that investigates the scientific extension and practical application of hazard energy within occupational safety [48].

5. Safety data collection, analysis, and sharing

A strong system of safety data collection, analysis, and sharing will assist the industry to understand the root causes of an event, explore existing and potential hazards, and continuously improving existing safety programs. Different countries and industries have conducted multiple reporting systems to collect, analyze, and share information with the public. For example, HSE has collected data on fatal injuries, nonfatal injuries, and ill health through the Labour Force Survey (LFS). The nonfatal injury and ill health estimates from the LFS are based on averages over 3 years. The fatal injuries data are collected based on RIDDOR (the Reporting of Injuries, Diseases, and Dangerous Occurrences Regulations) reports. In the United States, the Occupational Safety and Health Administration (OSHA) inspects the workplace to ensure compliance with minimum safety standards. If OSHA compliance officers find any violations on a site, they may issue a citation and a penalty. A company that had more than 10 employees at any time during the last calendar year must keep OSHA injury and illness records. Even if an employer is not required to keep injury and illness records, they are still required to report to OSHA within 8 hours any workplace incidents that result in death or the hospitalization of three or more employees. If there is a serious accident at a job site in which three or more workers are hospitalized or someone is killed, OSHA must be notified. OSHA will then investigate the accident.

5.1 Safety data collection

According to Section 3, safety incident data contains leading indicators (e.g. near misses) and lagging indicators (injures, illnesses, fatalities). Near-miss and incident reporting programs have been promoted and developed across high-risk industries [49]. OSHA requires employers to report all work-related fatalities and severe injuries according to OSHA Regulations (Standards—29 CFR 1904):

- All employers are required to notify OSHA when an employee is killed on the job or suffers a work-related hospitalization, amputation, or loss of an eye.

- A fatality must be reported within 8 hours.

- An in-patient hospitalization, amputation, or eye loss must be reported within 24 hours.

Besides the OSHA recordkeeping, many of the current companies in the high-hazard industries use daily reporting applications (apps) to collect the safety data from their project sites. However, those data are mainly used internally for company future development. **Figure 2** shows examples of safety data collected by the Bureau of Labor Statistics (BLS) and safety apps.

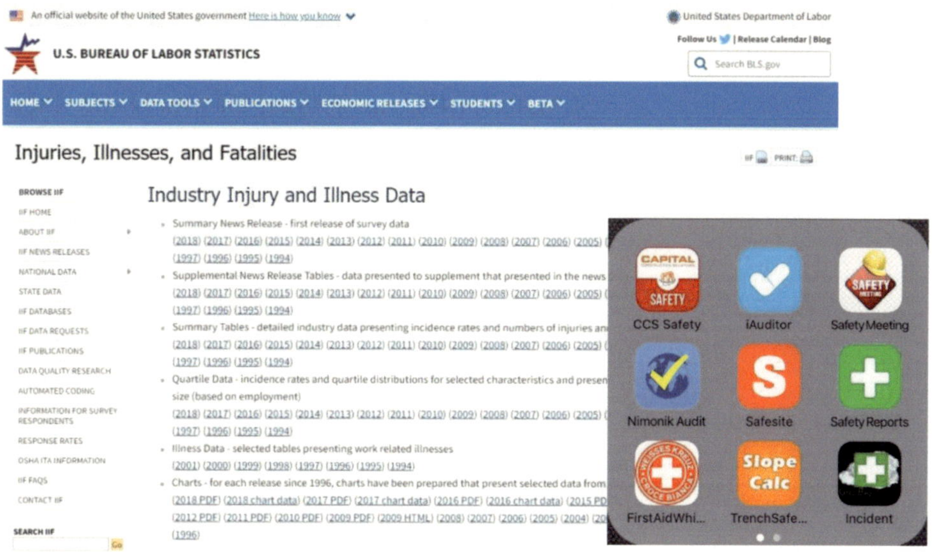

Figure 2.
Safety data collection methods (image source: https://www.bls.gov/iif/oshsum.htm and https://conappguru.com/apps/apps-for-construction-safety-2016/).

5.2 Safety data analysis

According to OSHA, the incidence rates represent the number of injuries and illnesses per 100 full-time workers and are calculated as $(N/\text{EH}) \times 200{,}000$:

$$\text{Incident rate} = \frac{(\#\text{of cases or days } per \text{ year}) \times 200{,}000}{\text{Total employee hours } per \text{ year}} \tag{1}$$

Where N = number of injuries and illnesses; EH = total hours worked by all employees during the calendar year; 200,000 = base for 100 equivalent full-time workers (working 40 hours per week, 50 weeks per year).

The US BLS Injuries, Illnesses, and Fatalities (IIF) program produces a wide range of information about workplace injuries and illnesses. These data are collected and reported annually through the Survey of Occupational Injuries and Illnesses (SOII) and the Census of Fatal Occupational Injuries (CFOI). **Table 2** is the latest industry incidence rates (OSHA recordable case rates) from BLS.

5.2.1 EMR (cost of accidents)

An accident cost usually includes direct and indirect costs. The biggest difference is if the costs can/cannot be directly attributed to the incident. The National Council on Compensation Insurance (NCCI)'s EMR is a metric to calculate workers' compensation insurance premiums.

EMR is calculated by trends in the loss ratio:

$$\text{Loss ratio} = \frac{\text{Claims (in \$)}}{\text{Premiums payments (in \$)}} \tag{2}$$

Industry	Total recordable cases		Cases with days away from work	
	2017	2018	2017	2018
Private industry	2.8	2.8	0.9	0.9
Agriculture, forestry, fishing, and hunting	5.0	5.3	1.7	1.7
Mining, quarrying, and oil and gas extraction	1.5	1.4	0.7	0.6
Construction	3.1	3.0	1.2	1.2
Manufacturing	2.5	3.4	0.9	0.9
Wholesale trade	2.8	2.9	1.0	1.0
Retail trade	3.3	3.5	1.0	1.1
Transportation and warehousing	4.6	4.5	2.0	2.1
Utilities	2.0	1.9	0.7	0.7
Information	1.3	1.3	0.6	0.6
Finance and insurance	0.5	0.5	0.1	0.1
Real estate and rental and leasing	2.4	2.3	1.0	0.8
Professional, scientific, and technical services	0.8	0.8	0.2	0.2
Management of companies and enterprises	0.9	0.8	0.2	0.2
Administrative and support and waste management and remediation services	2.2	2.3	0.9	0.9
Educational services	1.9	1.9	0.5	0.6
Health care and social assistance	4.1	3.9	1.1	1.1
Arts, entertainment, and recreation	4.2	4.1	1.2	1.1
Accommodation and food services	3.2	3.1	0.9	0.9
Other services (except public administration)	2.1	2.2	0.7	0.8

Table 2.
Incidence rates of nonfatal occupational injuries and illnesses by selected industry and case types, private industry, 2017-2018 [50].

The average EMR is 1.0. If a company's EMR is above 1.0, the company is considered riskier than most. For example, if a company has an EMR of 1.3, that means the insurance premiums could be up to 30% higher than a company with an EMR of 1.0. On the other hand, if a company has an EMR below 1.0, the company is considered safer than most which could receive a lower premium.

To better understand the collected incident data, many statistical modeling methods were used to identify the impact factors of incidents. The following two case studies introduced how statistical modeling helps with analyzing safety leading indicators and lagging indicators.

5.2.1.1 Case study #1 (using binary logit regression)

In this case study, approximately 2300 reported incidents at an active steel manufacturing facility in the U.S. between January of 2010 and August of 2016 were input into statistical predictive models [44]. The objective of this research is to identify specific variables using statistical models that increase the probability of an unsafe event or condition within a steel manufacturing facility [44]. Due to the organization and metrics recorded for the steel manufacturing safety incident database analyzed in this research, a statistical prediction model - Binary Logit Model was selected for data analysis. The probability denoted Pr(Y), is assumed to be determined by a set of independent variables (X_1, X_2, ..., X_j), and a corresponding set of parameters (β_0, β_1, β_2, ..., β_j). The dependent (Response) variables include OSHA Recordable, Lost time, First Aid, Property Damage, Environmental Incident, and Fire. The independent (predictor) variables describe the incident occurred situations [44]. Variables can be divided into six categories: summer indicator (June, July, or August), task performed indicators (operating or driving), moving equipment indicators (crane, truck, forklift, or trailer), mobile equipment indicator, location indicators (roll shop, coil yard/disposition, cut to length shop, west gate, water system, or melt shop), and preliminary cause indicators (defective equipment or personal responsibility) [44]. Findings from the regression analysis suggest that a positive correlation exists between incidents and summer months. One possible explanation is that employees have higher possibility to be distracted and fatigued due to high temperature. Results also suggest that injuries have a positive correlation with pedestrian employees near pieces of moving equipment [44]. Mobile equipment including trucks, forklifts, and truck and trailer combinations have a positive correlation with an incident [44].

5.2.1.2 Case study #2 (using text mining)

This study analyzes OSHA inspected fatalities data in the past 5 years from June 2014 to Aug 2018 with a total of 4769 accident records. Text mining techniques were deployed in this study for hazard report extraction [36]. **Figure 3** shows the research framework.

The incident description variables were processed using the R package 'openNLP' [51]. This package allows users to clean text data and perform machine-learning-based entity extractions. An energy source recognition method was then used to categorize the data into 10 energy source groups. Several corresponding key words were identified based on the energy source groups (**Table 3**). For example, "Worker died in fall from ladder" should be classified to Gravity due to the presence of the key term "fall". The findings show that gravity, motion, mechanical, and electrical related incidents have the largest percentage rate (**Figure 4**). This presented data analysis method can help with predicting future events, preventing reoccurrence of similar accidents, making scientific risk control plans, and incorporating hazard control measures into work tasks.

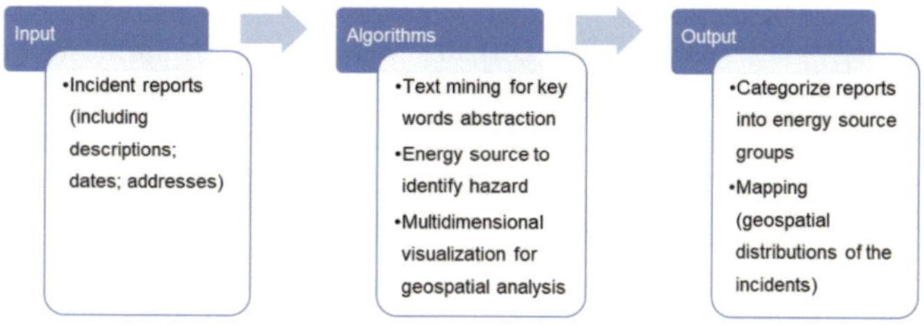

Figure 3.
Research framework of incident analysis using text mining and geospatial mapping [36].

Energy source	Corresponding key words
Gravity	Fall, excavation, collapsing, elevated, uneven, open holes
Motion	Confined space, movement, struck by, caught in, caught between, lifting
Mechanical	Rotating, compressed, conveyor, belt, motors, power tool, hand tool
Electrical	Electrocuted, power line, light fixtures, circuit panel, wiring, batteries
Pressure	Piping, cylinders, control liners, vessels, tanks, hoses, pneumatic, hydraulic
Temperature	Ignition, cold, hot, fumes, heat, molten slag
Chemical	Vapors, corrosive, gas, carbon monoxide, asphyxiation, chemical, toxic, sulfur dioxide
Biological	Animals, bacteria, viruses, insects, blood-borne pathogens, contaminated water, food
Radiation	Lighting, welding, arc, flash, X-rays, solar rays, microwaves, sunlight
Sound	Noise, vibration

Table 3.
Categorization of data into 10 energy source groups.

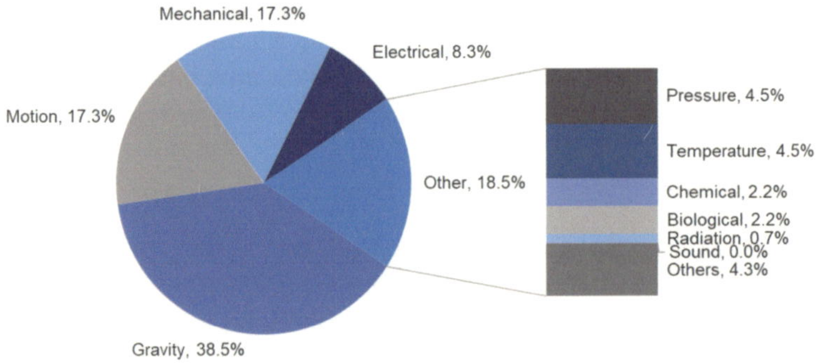

Figure 4.
Distribution of the energy sources [36].

6. Emerging technologies

6.1 Safety training through computer-aided technologies

The development of virtual reality (VR), augmented reality (AR), and mixed reality (MR) have embedded worker training systems and become significant cost-effective and safer ways to educate workers. The immersive VR/AR/MR environments within computer-generated simulations have also gained popularity in safety training to identify the potential hazards as well as educate moving vehicle operators on the job site. Hazardous construction scenarios can be simulated interactively with the working environment, workers' behavior, high-risk equipment, and working sequence [52]. Researchers also found that many VR/AR systems had been proved as efficient, usable, applicable, and accurate approaches in hazard identification, safety training, and education, and safety inspection [52].

6.2 Integrating BIM and safety

Numerous studies and industrial applications evidenced that safety and BIM integration can assist in safety planning and execution of projects, for example to automatic checking of construction models and schedules for preventing fall-related accidents; automated scaffolding-related safety hazard identification [53], visualization [54], and prevention [55], blind spots identification and mapping [56], path planning [57], near-miss information reporting and visualization [58], towel crane location optimization [59], etc.

6.3 Proximity detection devices

Many proximity avoidance systems have been developed by utilizing various technologies, such as an ultrasonic-based sensor [60], radio-frequency identification (RFID) sensing technology [61–63], radar [60, 64], GPS [65, 66], and magnetic field generators [67], to prevent contact accidents, particularly for accidents due to being struck by equipment. Most of these technologies provide some form of warning signals to workers when they are close to heavy equipment. These signals could be visual, vibratory, or audible warning signals [68].

6.4 Wearable sensing devices

A wide range of wearable devices has been applied across different industrial sectors including health care, manufacturing, mining, and athletics [69]. Some of these devices have proven to be very useful and beneficial to these industries and efforts are being made by both researchers and industry practitioners to improve on these technologies and learn from their initial implementation [70]. With the attention being gained by wearable devices worldwide, mobile devices are becoming part of everyday life and the number, types, and forms of wearable devices are increasing exponentially in recent years [71]. The most widespread adoption and implementation of wearable devices have been in the healthcare industry for the continuous monitoring of a user's physiological status [72]. For instance, wearable devices are used in the healthcare sector by patients personally to continuously monitor their physiological parameters and manage their health and well-being on a personal basis, or grant physicians remote access to their health data and receive personalized medical care [73]. Similarly, wearable devices incorporated with sensors such as the GPS, heart rate monitors, and pedometers are widely used in sports and fitness for tracking performance through unobtrusive and noninvasive monitoring and measurements [69]. Wearable sensors are integrated into a multitude of equipment used by professional athletes to monitor and measure their performance and safety [74]. For instance, sensors are incorporated into the helmets of National Football League (NFL) players to detect concussions and wired smart compression shirts are used to measure arm movement, and techniques are deployed to determine a pitcher's effectiveness in Major League Baseball (MLB). These different categories of wearable sensing devices can be efficiently deployed for safety and health data collection and analysis to provide real-time information to workers in industrial environments for accident prediction and prevention.

7. Conclusions

This chapter has discussed multiple industry safety-related topics including safety culture, hazard identification, safety leading and lagging indicators, safety data collection, analysis and sharing, and emerging technologies that can be embedded in safety management, training, and design. Multiple case studies and references were introduced to explain the different safety topics. Many more safety topics were not/briefly discussed in this chapter but still very important to know, for example, safety laws and regulations, design for safety, safety activity analysis, safety and productivity, heavy equipment management, occupational health illness-related topics, etc.

Author details

Siyuan Song[1*] and Ibukun Awolusi[2]

1 The University of Southern Mississippi, Hattiesburg, Mississippi, USA

2 The University of Texas at San Antonio, San Antonio, Texas, USA

*Address all correspondence to: siyuan.song@usm.edu

References

[1] Bureau of Labor Statistics (BLS). 2018 Census of Fatal Occupational Injuries (Final Data)—Industry by Event or Exposure. U.S. Bureau of Labor Statistics, U.S. Department of Labor; 2019. Available from: https://www.bls.gov/iif/oshwc/cfoi/cfch0016.pdf [Accessed: 31 July 2020]

[2] Health and Safety Executive (HSE). Construction Statistics in Great Britain, 2018. Health and Safety Executive. 2019. Available from: http://www.hse.gov.uk/statistics/industry/construction.pdf [Accessed: 31 July 2020]

[3] China Labour Bulletin (CLB). China's most Dangerous Industry is Getting more Dangerous. China Labour Bulletin. 2019. Available from: https://clb.org.hk/print/14884

[4] Hong Kong Occupational Safety and Health (HKOSH). Occupational Safety and Health Statistics 2017. Labor Department, The Government of the Hong Kong Special Administrative Region; 2017. Available from: https://www.labour.gov.hk/eng/osh/pdf/OSH_Statistics_2017_EN.pdf [Accessed: 31 July 2020]

[5] Verma A, Khan SD, Maiti J, Krishna OB. Identifying patterns of safety related incidents in a steel plant using association rule mining of incident investigation reports. Safety Science. 2014;**70**:89-98

[6] Reimink H. Safety and Health. 2016. Available from: http://www.worldsteel.org/publications/position-papers/safety-and-health.html [Accessed: 31 July 2020]

[7] Grabowski M, Ayyalasomayajula P, Merrick J, Harrald JR, Roberts K. Leading indicators of safety in virtual organizations. Safety Science. 2007;**45**(10):1013-1043. DOI: 10.1016/j.ssci.2006.09.007

[8] Choudhry RM, Fang D, Mohamed S. The nature of safety culture: A survey of the state-of-the-art. Safety Science. 2007;**45**(10):993-1012

[9] Hallowell MR, Hinze JW, Baud KC, Wehle A. Proactive construction safety control: Measuring, monitoring, and responding to safety leading indicators. Journal of Construction Engineering and Management. 2013;**139**(10):04013010

[10] Health and Safety Executive (HSE). Developing Process Safety Indicators: A Step-by-Step Guide for Chemical and Major Hazard Industries. London: Health and Safety Executive; 2006

[11] Awolusi I, Marks E. Near-Miss Reporting to Enhance Safety in the Steel Industry. Association for Iron and Steel Technology Publications: Safety First, Iron and Steel Technology; 2015. pp. 62-68

[12] Han S, Lee S. A vision-based motion capture and recognition framework for behavior-based safety management. Automation in Construction. 2013;**35**:131-141

[13] Advisory Committee on the Safety of Nuclear Installations (ACSNI). Study Group on Human Factors, Third Report: Organizing for Safety. London: HMSO; 1993

[14] Choudhry RM, Fang D, Mohamed S. The nature of safety culture: A survey of the state-of-the-art. Safety Science. 2006;**45**:993-1012

[15] Flannery JA. Safety Culture and its Measurement in Aviation. Newcastle: University of Newcastle Australia; 2001

[16] Fang D, Chen Y, Wong L. Safety climate in construction industry: A case study in Hong Kong. Journal of Construction Engineering and Management. 2006;**132**(6):573-584

[17] Cooper MD. Towards a model of safety culture. Safety Science. 2000;**36**(2):111-136

[18] Reason J, Hobbs A. Managing Maintenance Error: A Practical Guide. CRC Press; 2003

[19] Glendon AI, Stanton NA. Perspectives on safety culture. Safety Science. 2000;**34**(1-3):193-214

[20] Guldenmund FW. The nature of safety culture: A review of theory and research. Safety Science. 2000;**34**(1-3):215-257

[21] Wamuziri S. Safety culture in the construction industry. Proceedings of the Institution of Civil Engineers: Municipal Engineer. 2006;**159**(3):167-174

[22] Center for Offshore Safety (COS). Guidelines for a Robust Safety Culture. 2018. Available from: https://centerforoffshoresafety.org/~/media/COS/COSReboot/SEMS%20Good%20Practices/COS-3-04%20Guidelines%20for%20a%20Robust%20Safety%20Culture%20First%20Edition.pdf [Accessed: 31 July 2020]

[23] European Union Agency for Railways (ERA), Safety Culture. 2020. Available from: https://www.era.europa.eu/activities/safety-culture_en [Accessed: 31 July 2020]

[24] Abudayyeh O, Fredericks T, Butt S, Shaar A. An investigation of management's commitment to construction safety. International Journal of Project Management. 2006;**24**:167-174

[25] Song S, Awolusi I, Marks E. Impact of discretionary safety funding on construction safety. Journal of Safety Health and Environmental Research. 2017;**13**(2):378-384

[26] Chen Y, McCabe B, Hyatt D. A reslience safety climate model predicting construction safety performance. Safety Science. 2018;**109**:434-445

[27] Hinze J, Thurman S, Wehle A. Leading indicators of construction safety performance. Safety Science. 2013;**51**(1):23-28

[28] Awolusi IG, Marks ED. Safety activity analysis framework to evaluate safety performance in construction. Journal of Construction Engineering and Management. 2017;**143**(3):05016022

[29] Organization for Economic Cooperation and Development (OECD). Guidance on Safety Performance Indicators. OECD Environment. Series on Chemical Accidents No. 11. Health and Safety Publications; 2003

[30] Reiman T, Pietikäinen E. Leading indicators of system safety—monitoring and driving the organizational safety potential. Safety Science. 2012;**50**(10):1993-2000

[31] Awolusi I, Nnaji C, Marks E, Hallowell M. Enhancing construction safety monitoring through the application of internet of things and wearable sensing devices: A review. In: Computing in Civil Engineering 2019: Data, Sensing, and Analytics. Reston, VA: American Society of Civil Engineers; 2019. pp. 530-538

[32] Oien K, Utne IB, Herrera IA. Building safety indicators: Part 1—Theoretical foundation. Safety Science. 2011;**49**(2):148-161

[33] Pradhananga N, Teizer J. Automatic spatio-temporal analysis of construction site equipment operations using GPS data. Automation in Construction. 2013;**29**:107-122

[34] Hinze J, Madariaga M, Pizarro D. Target Safety: Programs Focused on Preventing Specific Hazards.

Construction Industry Institute; 2006. pp. 216-211

[35] Alruqi WM, Hallowell MR. Critical success factors for construction safety: Review and meta-analysis of safety leading indicators. Journal of Construction Engineering and Management. 2019;**145**(3):04019005

[36] Song S, Awolusi I, Jiang Z. Work-Related Fatalities Analysis Through Energy Source Recognition. Tempe, AZ: Construction Research Congress; 2020

[37] McDonald MA, Lipscomb HJ, Bondy J, Glazner J. Safety is everyone's job: The key to safety on a large university construction site. Journal of Safety Research. 2009;**40**(1):53-61

[38] Albert A, Hallowell MR. Hazard recognition methods in the construction industry. Construction Research Congress. 2012;**2012**:407-416

[39] Bobick TG. Falls through roof and floor openings and surfaces, including skylights: 1992-2000. Journal of Construction Engineering Management. 2004;**130**(6):895-907

[40] Tixier AJ, Albert A, Hallowell MR. Teaching construction hazard recognition through high fidelity augmented reality. In: 120th ASEE Annual Conference and Exposition; 23-26 June 2013. 2013

[41] Han S, Yang H, Im D. Designing a human-computer interface for a process control room: A case study of a steel manufacturing company. International Journal of Industrial Ergonomics. 2007;**37**(5):383-393

[42] Dhar N, Kamruzzaman M, Ahmed M. Effect of minimum quantity lubrication (MQL) on tool wear and surface roughness in turning AISI-4340 steel. Journal of Materials Processing Technology. 2006;**172**(2):299-304

[43] Ologe F, Akande T, Olajide T. Occupational noise exposure and sensorineural hearing loss among workers of a steel rolling mill. European Archives of Oto-Rhino-Laryngology. 2006;**263**(7):618-621

[44] Song S, Lyu Q, Marks E, Hainen A. Steel manufacturing incident analysis and prediction. Journal of Safety, Health and Environmental Research. 2018;**14**(1):331-336

[45] Tixier AJ, Albert A, Hallowell MR. Proposing and validating a new way of construction Hazard recognition training in academia: Mixed-method approach. Practice Periodical on Structural Design and Construction. 2018;**23**(1):1-10

[46] Behm M, Schneller A. Application of the Loughborough construction accident causation model: A framework for organizational learning. Construction Management and Economics. 2013;**31**(6):580-595

[47] Goh YM, Chua DKH. Case-based reasoning approach to construction safety Hazard identification: Adaptation and utilization. Journal of Construction Engineering and Management. 2010;**136**(2):170-178

[48] Hallowell MR, Alexander D, Gambatese JA. Energy-based safety risk assessment: Does magnitude and intensity of energy predict injury severity? Construction Management and Economics. 2017;**35**(1-2):64-77

[49] Barach P, Small SD. Reporting and preventing medical mishaps: Lessons from non-medical near miss reporting systems. BMJ. 2000;**320**(759-763):2000

[50] Bureau of Labor Statistics (BLS). Employer-Reported Workplace Injuries and Illnesses. 2018. Available from: https://www.bls.gov/news.release/archives/osh_11072019.pdf [Accessed: 31 July 2020]

[51] Baldridge J, Morton T, Bierner G. The openNLP MAXENT package. 2008. Available from: http://maxent.sourceforge.net

[52] Li X, Yi W, Chi HL, Wang X, Chan AP. A critical review of virtual and augmented reality (VR/AR) applications in construction safety. Automation in Construction. 2018;**86**:150-162

[53] Zhang S, Lee JK, Venugopal M, Teizer J, Eastman C. Integrating BIM and safety: An automated rule-based checking system for safety planning and simulation. Proceedings of CIB. 2011;**W099**(99):24-26

[54] Zhang S, Teizer J, Lee JK, Eastman CM, Venugopal M. Building information modeling (BIM) and safety: Automatic safety checking of construction models and schedules. Automation in Construction. 2013;**29**:183-195

[55] Kim K, Cho Y, Zhang S. Integrating work sequences and temporary structures into safety planning: Automated scaffolding-related safety hazard identification and prevention in BIM. Automation in Construction. 2016;**70**:128-142

[56] Song S, Marks E, Moynihan G. Dynamic 3D blind spot mapping for equipment operations. Journal of Safety, Health and Environmental Research. 2019;**15**(1):348-353

[57] Song S, Marks E. Construction site path planning optimization through BIM. In: Computing in Civil Engineering 2019: Visualization, Information Modeling, and Simulation. Reston, VA: American Society of Civil Engineers; 2019. pp. 369-376

[58] Shen X, Marks E. Near-miss information visualization tool in BIM for construction safety. Journal of Construction Engineering and Management. 2016;**142**(4):04015100

[59] Shen X, Marks E. Site location optimization of a tower crane through building information modeling (BIM). Journal of Safety, Health and Environmental Research. 2017;**13**:330-337

[60] Choe S, Leite F, Seedah D, Caldas C. Evaluation of sensing technology for the prevention of backover accidents in construction work zones. Journal of Information Technology in Construction. 2014;**19**(August 2013):1-19

[61] Chae S, Yoshida T. Automation in construction application of RFID technology to prevention of collision accident with heavy equipment. Automation in Construction. 2010;**19**(3):368-374. DOI: 10.1016/j.autcon.2009.12.008

[62] Teizer J, Allread BS, Fullerton CE, Hinze J. Autonomous pro-active real-time construction worker and equipment operator proximity safety alert system. Automation in Construction. 2010;**19**(5):630-640

[63] Park J, Marks E, Cho YK, Suryanto W. Performance test of wireless technologies for personnel and equipment proximity sensing in work zones. Journal of Construction Engineering and Management. 2016;**142**(1):04015049-1-04015049-9

[64] Ruff TM, Holden TP. Preventing collisions involving surface mining equipment: A GPS-based approach. Journal of Safety Research. 2003;**34**(2):175-181

[65] Oloufa AA, Ikeda M, Oda H. Situational awareness of construction equipment using GPS, wireless and web technologies. Automation in Construction. 2003;**12**:737-748

[66] Wang J, Razavi SN. Low false alarm rate model for unsafe-proximity detection in construction. Journal of Computing in Civil Engineering. 2016;**30**(2):1-13

[67] Li J, Carr J, Jobes C. A shell-based magnetic field model for magnetic proximity detection systems. Safety Science. 2012;**50**(3):463-471

[68] Awolusi I, Song S, Marks E. Forklift safety: Sensing the dangers with technology. Professional Safety. 2017;**62**(10):36-39

[69] Awolusi I, Nnaji C, Okpala I. Success Factors for the Implementation of Wearable Sensing Devices for Safety and Health Monitoring in Construction. Tempe, AZ: Construction Research Congress; 2020

[70] Anzaldo D. Wearable sports technology—market landscape and compute SoC trends. In: ISOCC 2015—International SoC Design Conference: SoC for Internet of Everything (IoE); 1-6 November 2015. 2015

[71] Nnaji C, Okpala I, Awolusi I. Wearable sensing devices: Potential impact and current use for incident prevention. Professional Safety. 2020;**65**(04):16-24

[72] Led S, Azpilicueta L, Martínez-Espronceda M, Serrano L, Falcone F. Operation, analysis and optimization of wireless sensor devices in health-oriented monitoring systems. Mobile Health. 2015:245-263

[73] Marakhimov A, Joo J. Consumer adaptation and infusion of wearable devices for healthcare. Computers in Human Behavior. 2017;**76**:135-148

[74] Guta M. The Impact of Wearable Tech in Sports. Wearable Tech World. 2014. Available from: http://www.wearabletechworld.com/topics/wearable-tech/articles/376634-impact-wearable-tech-sports.htm

Chapter 6

Integrated Lean-Green-Six Sigma Practices to Improve the Performance of the Manufacturing Industry

Lokpriya M. Gaikwad and Vivek K. Sunnapwar

Abstract

To survive in the global competitive market, the manufacturing organization must adopt changes in technologies and strategies into their processes on a continuous basis. So, nowadays, Lean, Green, and Six Sigma became business process strategies, which are employed in most of the organization to enhance their manufacturing performance. However, the significant information is that these strategies are implemented sequentially instead of simultaneously. The objective of this chapter is to propose an integrated Lean- Green-Six Sigma strategic framework for manufacturing industries that effectively implementing this approach will lead business processes to achieve operational, financial, social, and environmental growth. This will also guide the practitioner and academician those who are working on manufacturing strategies.

Keywords: Lean practices, Green practices, Six Sigma, integrated framework

1. Introduction

The expanded mindfulness about sustainability and requests for eco-friendly products has constrained the manufacturing firms to reevaluate their business activities [1]. Ventures are embracing the conventional strategies for manufacturing and petroleum products in the majority of the world [1]. The manufacturing enterprises in developed countries discharge multiple times carbon dioxide when contrasted with the developing nations [2]. The current arrangements on environmental change uncover that the normal surface temperature of the earth will increase to 3°C until the finish of this century [2]. This temperature increment is far away from the aggressive objective of the Paris Agreement that plans to confine it up to 2°C [3].

The hazard related to the atmosphere will increase in the offing because of expanded risk recurrence and increasingly powerless populace. Subsequently, for the government assistance of society and environmental assurance, enterprises must remember green innovations for their tasks. The presence of present-day businesses relies on their fitness to change with the external environment [4]. In this way, to stay practical in the market, manufacturing firms must create and actualize low carbon emanation advancements. Thus, the firms are spending huge money to devise sustainable techniques for creation and utilization [5].

Since the most recent couple of decades, numerous thoughts and approaches have been created such as Lean, Green, Six Sigma, and so forth [6, 7] to create the top quality items. Gaikwad and Sunnapwar [8] introduced critical achievement factors for executing incorporated LGSS rehearses in the manufacturing industry.

Yet, an individual methodology cannot address all the issues comprehensively identified with sustainability [9]. Along these lines, a coordinated methodology is required that lessens wastes and varieties and mitigates negative ecological effects [3]. Sreedharan and Raju [10] opined that GLS (Green Lean Six Sigma) is a comprehensive methodology that reduces ecological outflows through 3R's (reduce, reuse, and recycle).

LGSS comprises of three remarkable methodologies viz., Green, Lean, and Six Sigma that expand profitability elements through reduced emissions, wastes, and rework [11]. Lean promotes the efficient evacuation of wastes through excellence at all levels inside the firm [12, 13]. Green innovation diminishes the negative ecological effect of the item by making it all the more ecofriendly [14–16]. Six Sigma decreases varieties in the process that prompts diminished dismissal of products [17]. However, consolidated Green Lean Six Sigma is fit for creating an item that is not just of high caliber and ease yet in addition eco-friendly [18].

In the literature, there is significant proof of the combination and system of Lean and Green, Lean, and Six Sigma. In any case, the literature needs enough proof of the integration and structure of LGSS. There exist hardly any investigations related to the LGSS structure; however, the concerned associations confronted difficulties to execute the equivalent because of the conventional nature of the system [19]. In addition, in the literature, there exists no examination on the LGSS structure that can be applied independently of the size, type, and culture of the firm. For this, the current study gives reconciliation and system of LGSS for accomplishing greatness in profitability and environmental sustainability.

Lean Six Sigma is an approach that identifies the customer expectations, eliminates wastages, and reduces variability. It combines Six Sigma's problem-solving approach with statistical tools with a Lean effect on flow improvement [20]. Hence, the objective of industrial engineering to improve productivity through optimization techniques has been fulfilled.

Evidence suggests that LGSS approaches make a positive impact on social, economical, and environmental (i.e., sustainability) performance of the firm [21]. Therefore, industrial engineering practitioners make an effort toward the deployment of such types of approaches into their business. The industrial engineering field has also tremendous pressure to improve production efficiency concerning social and environmental responsibility, and also market dynamics have changed [15, 16]. Bergmiller and McCright [22] opined that the three elements of sustainability (economic, social, and environmental) should be mulled over by firms to keep their competitive edge. In this situation, the confront for firms is to meet all their stakeholders' requirements through achieving positive financial execution while finding the correct equilibrium among the triple bottom line of sustainability [23, 24].

2. Literature review

The historical backdrop of GLS followed back to the advancement of Lean way of the thinking. The Lean idea was contrived in Japan after the Universal War II to contend with the large scale manufacturing arrangement of the USA [5, 25]. The

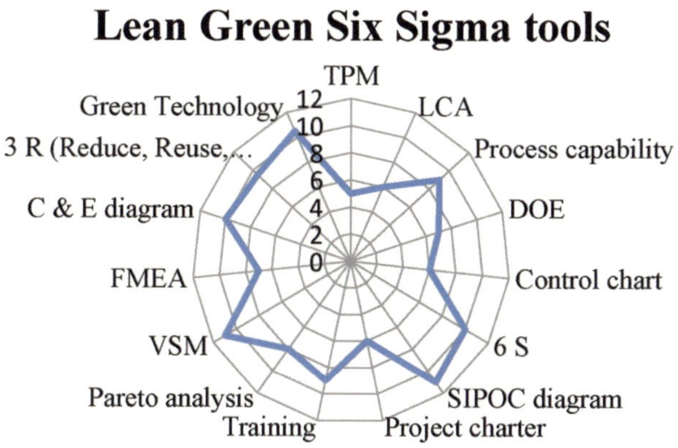

Figure 1.
Radar chart of LGSS tools.

cutting edge idea of Lean philosophy originated from the Toyota Production System (TPS), spearheaded by Japanese specialists Taiichi Ohno and Shigeo Shingo [3]. The Lean methodology diminishes wastes; it cannot lessen the unfriendly natural effects [26].

The execution of Green innovations can defeat this confinement of the Lean. Green innovation adds to Lean as it diminishes negative natural effects and other related wastes [27]. Green innovation decreases natural discharges; it cannot lessen the lean wastes [1]. Along these lines, there is a requirement for a consolidated Green Lean (GL) approach that limits Lean wastes as well as decreases the carbon footprint. There is a decent arrangement of similitude between the two strategies dependent on waste decrease procedures and management practices [22].

The majority of the investigations discover that there is an absence of metrics, top management backing, and standard method for execution of LGSS in the firm. The LGSS instruments can possibly discover the wastages, abandons, and undesirable exercises that lead to the advancement of resources [28]. As suggested by Kaswan and Rathi [4], LGSS tools are presented in **Figure 1** in the form of a radar chart.

3. Integration of lean green six sigma

The GLS has gotten due consideration as of late due to its capacity to improve efficiency, and profitability and alleviate environmental concerns [19]. LGSS framework proposes a hypothetical incorporation model of the LGSS dependent on the blend of hypothetical components. **Figure 2** delineates a coordinated LGSS framework. The reconciliation model's principle point is to depict the basic realities required for the manufacturing industry to improve sustainable execution. The proposed model speaks to the theoretical likenesses between the three methodologies. The empowering influences function as the key data sources that invigorate the incorporation of LGSS while the exhibitions in trade-off fill in as yield. The difficulties for the LGSS combination are the limitations that confine the hierarchical interests to improve the manageability elements. The apparatuses and related LGSS framework are considered as the supporting mechanism that bolsters the joining and execution of LGSS.

The goal is to maximize stakeholder value, by achieving customer satisfaction, process speed, cost, and quality through the tools and techniques [29]. This

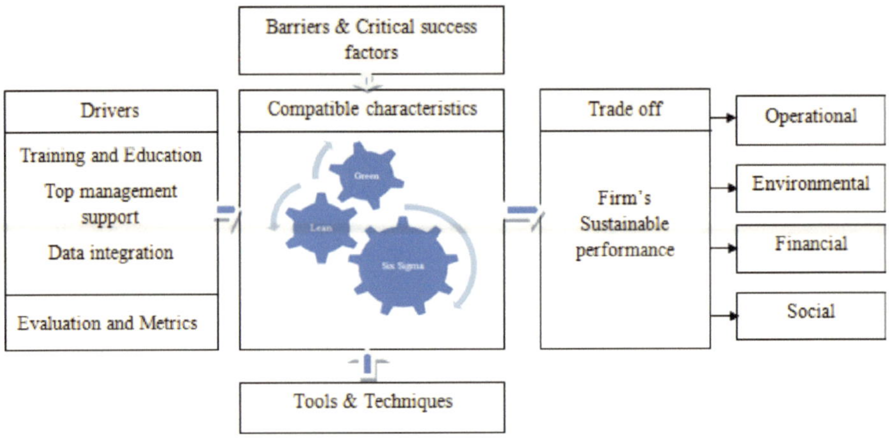

Figure 2.
Integrated framework of LGSS.

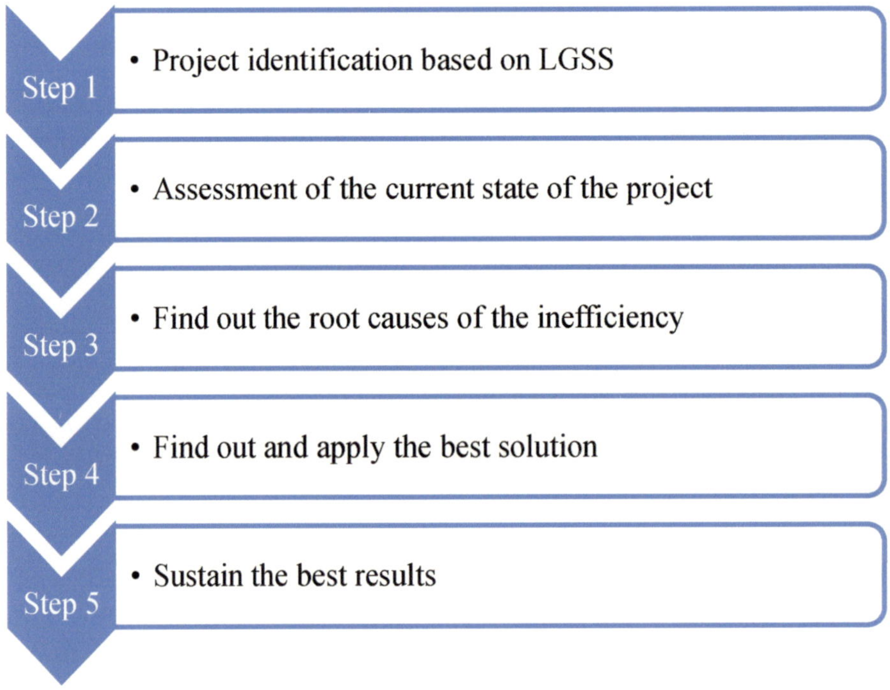

Figure 3.
Execution steps of LGSS framework.

initiative depends upon collaborative team effort to improve performance of the firm [30] to identify the need of the customers and to eliminate unwanted activities within the process [31]. The individual LGSS approach shares some similarities such as employee involvement, customer satisfaction, and continuous improvement [32]. Aqlan and Alfandi [33] opined that each of the practice has been carried out by many forms to improve their processes. Some researchers focus on Lean achievement in excellence through the reduction of waste and unwanted activities [34].

Following **Figure 3** depicts the Six Sigma DMAIC (Define-Measure-Analyze-Improve-Control)-based framework that can be adopted by all business firms. The proposed integrated framework has executed in the following five steps.

3.1 Execution steps of LGSS framework

3.1.1 Step 1: project identification based on LGSS

In this step, the project may be selected based on wastage, defect rate, environmental issues faced by the organization, social issues, financial issues, and voice of the customers. The selected project can be classified into different departments/sections of the company that is initiated the LGSS project. Gupta et al. [35] stated that most of the projects fail due to improper selection. Selection of the LGSS project depends upon sustainability need and global competitive pressure that force the firm to select the project. So, based on the customer demand, environmental concern, and socio-economic need, the final project is selected. After identification of a feasible project, a charter is prepared based on the scope of the project, schedule, and team members that involved in the project.

3.1.2 Step 2: assessment of the current state of the project

In this step, the current state of nature of the selected project has been estimated. Here, the performance of the selected project is compared with the indices of Lean, Green, and Six Sigma practices. Based on the collected data, the std. deviation, sigma level, and C_{pk} can be estimated using statistical tools. Estimations such as CO_2 emission, material consumption, Green energy coefficient are estimated using green energy technology tools like life cycle assessment. VSM is the best tool of Lean practices for evaluating the current degree of waste in the process. The blend of significant VSM and life cycle assessment lead to the evaluation of Lean-Green wastes that gives the source to assist improvement.

3.1.3 Step 3: find out the root causes of the inefficiency

In this progression, the underlying drivers of the wastefulness identified with wastages, emanation, and imperfection rate in the task have been distinguished. From the client and business perspective, the worth included and nonvalue included activities have been distinguished, and afterward, the procedure cycle effectiveness is resolved to contrast with world-class benchmarking value which decides how much improvement is required.

In the interim, the imperatives and bottlenecks are discovered in the chosen project and investigations identified with deformity rate, discharge rate, and varieties in the process are discovered. Advanced measurable quality tools, such as fishbone diagram, 5 Why, FMEA, and so on, are utilized to discover the main driver of the issue.

3.1.4 Step 4: find out and apply the best solution

Once the causes of inefficiency and wastages are find out, potential solutions are proposed and tested, and the best solution is applied to diminish the prominent reasons. All the sources of information such as customers, staff, stakeholders, project sponsors are used to determine the evaluation criteria. To evaluate the proposed solution against the criteria, tools such as pug matrix, DOE (Design of experiment), LCA, etc. are used. The estimation of time-saving, improved quality, and other quality measures is also made with enhanced VSM. After launching the best solution as a pilot solution, the task to be performed is documented, and pilot participants are trained in various aspects of the best solution. After that, the pilot solution is executed in a selected section or department of the company.

3.1.5 Step 5: sustain the best results

In this step, the results are sustained in the process or system if the considerable improvement is observed. The whole process is reevaluated using lifecycle assessment (LCA) and value stream mapping (VSM) techniques to judge the level of emissions and waste reduction. Also, various observations, control charts, and data collection techniques are used to reassess the sigma level, $C_{pk,}$ electricity, water, material consumption, etc. If the sustained results are not found, then an action plan is also initiated to select an appropriate solution. Through the execution of LGSS, the firm can gain improved sustainability, reputation, and global market through the delivery of eco-friendly products.

4. Results and discussion

LGSS is in the evolution phase and most of the firms oppose adopting such type of strategy due to the change in routine work method into their operation. However, the presented framework will help the organization for improving sustainable performance through overcoming the barriers and constraints which come in front of the execution.

After the integration of these strategies, one can correlate the different functionality of these modern techniques. It has found that top management support, a culture of the organization, and team work are the crucial success factors for the execution of such a program [4, 8]. In the same way, there are certain challenges to adopt these strategies because of improper guidance and roadmap to achieve set objectives of the firm [36]. There is a need to find out such types of obstacles during the execution of the improvement program within a given time frame to take a competitive advantage.

The LGSS tools and techniques overcome the barriers and at the same type advanced statistical tools help to find out the root cause of the problems and also to find out the solution for such type of problems. Currently, most of the organizations want to implement such types of manufacturing philosophies into their operation to enhance productivity, on-time delivery, improved quality, and reduced cost. In the LGSS framework, the Define-Measure-Analyze-Improve-Control (DMAIC) approach helps to overcome the barriers of the Lean-Green approach to delivering the sustainable results.

5. Proposed plan for validation

"The limitation of this work is that the presented framework has not tested statistically; in the future, the research will be based on an empirical study to be carried out in different industrial sectors." To validate the proposed plan, an empirical study taking a survey of manufacturing industries in which those practices are going on into their operation has need to be taken and then it should be statistically validated using SPSS-AMOS software.

In execution steps as shown in **Figure 3**, problem identification should be based on LGSS practices. Meanwhile, every problem that faces in the industry must be identified and solved by LGSS tools and techniques. Moving toward step 2, there should be an assessment of the current state of the problem to know the issues in the process. After that, the main drivers of the issues are associated with the inefficiency. Hence, the root causes using the fishbone diagram, 5 Why, DOE, etc. should be found out to deliver productivity to the system. After finding the root cause of the problem, the best solution for implementation into practices should be found out. However, the best solution for the implementation should be sustained for continuous improvement. Everyone within the organization should be capable of finding solutions for the problems and sustaining the best solution into practice.

6. Conclusion

The LGSS approach reduces the negative environmental effect and improves operational, social, and financial performance. The integrated LGSS framework presented is based on theoretical elements such as tools and techniques, barriers, and enablers. Top management support and teamwork have a significant effect on execution of LGSS. The proposed framework provides a way of starting from project identification to finding a solution to the problem of implementation in a system/process. The stepwise framework of LGSS provides help to the operation manager to execute this sustainable approach irrespective of the size, culture, and type of industry. Providing this integrated approach, definitely, the organization can enhance its operational productivity keeping in mind socio-economic needs and global competitive pressure that force the organization to implement such type of approach.

Acknowledgements

The authors sincerely thank the editor and the anonymous reviewers for their constructive and important comments on the chapter.

Conflict of interest

I confirm there is no "conflict of interest."

Author details

Lokpriya M. Gaikwad[1*] and Vivek K. Sunnapwar[2]

1 Mechanical Engineering Department, Sardar Patel College of Engineering, Mumbai, India

2 Lokmanya Tilak College of Engineering, Navi Mumbai, India

*Address all correspondence to: lokpriya2007@gmail.com

References

[1] Siegel R, Antony J, Garza-Reyes JA, Cherrafi A, Lameijer B. Integrated green lean approach and sustainability for SMEs: From a literature review to a conceptual framework. Journal of Cleaner Production. 2019;**240**:118205

[2] Intergovernmental Panel on Climate Change. Global warming of 1.5° C: An IPCC special report on the impacts of global warming of 1.5° C above pre-industrial levels and related global greenhouse gas emission pathways. In: The Context of Strengthening the Global Response to the Threat of Climate Change, Sustainable Development, and Efforts to Eradicate Poverty. Incheon, Republic of Korea: Intergovernmental Panel on Climate Change; 2018

[3] Kaswan MS, Rathi R. Analysis and modeling the enablers of green lean six sigma implementation using interpretive structural modeling. Journal of Cleaner Production. 2019;**231**:1182-1191

[4] Kaswan MS, Rathi R. Green lean six sigma for sustainable development: Integration and framework. Environmental impact assessment review. 2020;**83**:106396

[5] Duarte S, Cruz-Machado V. Green and lean supply-chain transformation: A roadmap. Production Planning and Control. 2019:1-13

[6] Bhat S, Gijo EV, Jnanesh NA. Application of lean six sigma methodology in the registration process of a hospital. International Journal of Productivity and Performance Management. 2014;**63**(5):613-643

[7] Garza-Reyes JA, Yu M, Kumar V, Upadhyay A. Total quality environmental management: Adoption status in the Chinese manufacturing sector. TQM Journal. 2018;**30**(1):2-19

[8] Gaikwad LM, Sunnapwar V. Critical success factors for implementing integrated lean green six sigma strategies: A framework. Journal of Industrial Engineering and its Applications. 2020a;**5**(1):25-40

[9] Pandey H, Garg D, Luthra S. Identification and ranking of enablers of green lean six sigma implementation using AHP. International Journal of Production Quality Management. 2018;**23**(2):187-217

[10] Sreedharan VR, Raju R. A systematic literature review of lean six sigma in different industries. International journal of Lean Six Sigma. 2016;7(4):430-466

[11] Kumar S, Luthra S, Govindan K, Kumar N, Haleem A. Barriers in green lean six sigma product development process: An ISM approach. Production Planning and Control. 2016;**27**(7-8):604-620

[12] Panwar A, Jain R, Rathore APS, Nepal B, Lyons AC. The impact of lean practices on operational performance–an empirical investigation of Indian process industries. Production Planning and Control. 2018;**29**(2):158-169

[13] Zhu Q, Johnson S, Sarkis J. Lean six sigma and environmental sustainability: A hospital perspective. Supply Chain Forum: An International Journal. 2018;**19**(1):25-41

[14] Deif AM. A system model for green manufacturing. Journal of Cleaner Production. 2011;**19**(14):1553-1559

[15] Garza-Reyes JA. Lean and green—A systematic review of the state of the art literature. Journal of Cleaner Production. 2015;**102**:18-29

[16] Garza-Reyes JA. Green lean and the need for six sigma. International Journal of Lean Six Sigma. 2015;**6**(3):226-248

[17] Kumaravadivel A, Natarajan U. Application of six-sigma DMAIC methodology to sand-casting process with response surface methodology. International Journal of Advanced Manufacturing Technology. 2013;**69**(5-8):1403-1420

[18] Hussain K, He Z, Ahmad N, Iqbal M. Green, lean, six sigma barriers at a glance: A case from the construction sector of Pakistan. Building and Environment. 2019;**106225**

[19] Sony M, Naik S. Green lean six sigma implementation framework: A case of reducing graphite and dust pollution. International Journal of Sustainable Engineering. 2019;**1-10**

[20] Message Costa LB, Filho MG, Fredendall LD, Devós Ganga GM. Lean six sigma in the food industry: Construct development and measurement validation. International Journal of Production Economics. 2020;**231**. DOI: 10.1016/j.ijpe.2020.107843

[21] Cherrafi A, Elfezazi S, Govindan K, Garza-Reyes JA, Benhida K, Mokhlis A. A framework for the integration of green and lean six sigma for superior sustainability performance. International Journal of Production Research. 2016;**55**. DOI: 10.1080/00207543.2016.1266406

[22] Bergmiller GG, McCright PR. Parallel models for lean and green operations. In: Proceedings of the 2009 Industrial Engineering Research Conference. Florida: The University of South Florida and Zero Waste Operations Research and Consulting; 2009. pp. 1138-1143

[23] Alves JRX, Alves JM. Production management model integrating the principles of lean manufacturing and sustainability supported by the cultural transformation of a company. International Journal of Production Research. 2015;**51**(11):1-14

[24] McCarty T, Jordan M, Probst D. Six Sigma for Sustainability–How Organizations Design and Deploy Winning Environmental Programs. New York: McGraw–Hill; 2011

[25] Bhamu J, Singh Sangwan K. Lean manufacturing: Literature review and research issues. International Journal of Operations & Production Management. 2014;**34**(7):876-940

[26] Aggogeri F. Lean thinking to change healthcare organizations: A case study to reduce waste and redesign services. European Journal of Cross-Cultural Competence and Management. 2014;**3**(3-4):196-211

[27] Dües CM, Tan KH, Lim M. Green as the new lean: How to use lean practices as a catalyst to greening your supply chain. J. Cleaner Production. 2013;**40**:93-100

[28] Gaikwad LM, Sunnapwar V. An integrated approach for lean, green and six sigma strategies: A systematic literature review and future directions for developing a specific framework. TQM Journal. 2020b;**32**(2):201-222. DOI: 10.1108/TQM-08-2018-0114

[29] Laureani A, Antony J. Standards for lean six sigma certification. International Journal of Productivity and Performance Management. 2011;**61**:110-120. DOI: 10.1108/17410401211188560

[30] Rajak AK, Niraj M, Kumar S. In pursuit of lean six sigma: A systematic review. International Journal of Applied Engineering Research. 2016;**11**:547-556

[31] Yadav G, Desai TN. Lean six sigma: A categorized review of the literature. International Journal of Lean Six Sigma. 2016;7:2-24. DOI: 10.1108/IJLSS-05-2015-0015

[32] Laureani A, Antony J. Leadership characteristics for lean six sigma. Total

Quality Management. 2017;**28**:405-426. DOI: 10.1108/17410401211188560

[33] Aqlan F, Al-fandi L. Prioritizing process improvement initiatives in manufacturing environments. International Journal of Production Economics. 2018;**196**:261-268

[34] Fullerton RR, Kennedy FA, Widener SK, Huntsman JM. Lean manufacturing and firm performance: The incremental contribution of lean management accounting practices. Journal of Operations Management. 2014;**32**:414-428

[35] Gupta SK, Gunasekaran A, Antony J, Gupta S, Bag S, Roubaud D. Systematic literature review of project failures: Current trends and scope for future research. Computers and Industrial Engineering. 2019;**127**:274-285

[36] Gamal Aboelmaged M. Reconstructing six sigma barriers in manufacturing and service organizations: The effects of organizational parameters. International Journal of Quality & Reliability Management. 2011;**28**(5):519-541

Section 3

Total Systems Approach to Operational Decision-Making

Chapter 7

A Hybrid Human-Data Methodology for the Conception of Operational Performance Management Systems

Diogo Ginjo Jantarada and Antonio Grilo

Abstract

To excel in the overall business performance, daily processes and activities connected to produce a good or service need to be outperformed. Even though there is extensive literature on performance management and performance management systems, there is still no consensus over the conceptual model of such systems, in what is designated as Operational Performance Management Systems (OPMS). This chapter proposes a new approach to conceive feasible and desirable OPMS tools to assist managers on controlling and responding to operational needs, by combining Design Thinking (DT) and Data Analytics (DA), that provide holistic and deep business knowledge, as well as a data-driven based management. The authors conduct an empirical application through a case study within the context of a European airport's Baggage Handling System (BHS). The case study procedure follows the proposed methodology's stages, where the authors construct the problem space with a wide array of collected data, along with the solution exploration and refinement.

Keywords: operational performance management, operational performance management systems, design thinking, data analytics, conception, design

1. Introduction

With the rise of the modern society and its socio-economic structure based on companies (public or private, profit or non-profit based) and its interdependences, the need for performance excellence has become of the utmost importance. The twenty-first century marked the beginning of a new millennium where uncertainty is at its high and companies are constantly trying to outstand their business performance to thrive and gain competitive advantage.

To excel on their overall business performance, the same must apply to the operational performance. Thus, operational performance management systems must reflect the problems and impediments hindering the company's competitiveness. The trends show they are trying to gain knowledge about themselves and the market by leveraging ICT and Data Analytics (DA) capabilities to transform data into information [1, 2]. At the same time, they are focused on understanding the user's (internal or external) pains and needs and come up with a way of meeting them to improve business performance [3, 4].

This double-sided story shows us both a hard part (described from an objective transformation of the real world into data) and soft part (where humans and their different perspectives of the real world are considered). However, there is little research combining 'soft' and 'hard' methodologies to tackle problems or develop new products or services which gather the emphatic human-centric approach with the rigorous and objective perspective of data science.

Thus, the author's objective is to propose a new hybrid methodology which uses a Design Thinking approach to an unknown problem environment and develops analytic-based solutions to create OPMS tools. This is intended to provide:

- the full picture of the operational context within a multi-stakeholder environment;
- a broad but clear definition of the problem-space to have a common basis for ideation within a collaborative environment;
- digital solutions for managers to control effectively their operations;
- confirm DT outputs which might seem too intuitive and subjective to traditional companies.

The proposed methodology was tested under real-life conditions by performing a case study investigation methodology on a Baggage Handling System (BHS) company, within a European airport. This allowed to test various tools and the overall procedure and is able to critically assess the advantages and limitations of this methodology for the intended purposes.

2. Methodology foundations

This chapter draws upon three main research topics: Operational Performance Management, Design Thinking and Data Analytics. These topics are briefly addressed hereinafter, as they lay the foundation for the proposed methodology.

2.1 Operational performance management

Drawing upon different Performance Management (PM) literature reviews [5–8], this subject area has been progressively developed throughout the twentieth and twenty-first centuries. Its evolution can be summarized into four major periods. The progressive phases had different objectives and perspectives over what issues should PM address. They constructed the current understanding of an ever-changing and multi-dimensional performance management discipline, with applications on a wider variety of contexts (e.g. small vs. large organizational dimensions, fixed-teams vs. collaborative environments, operational vs. marketing performance) and with a broader scope of objectives (e.g. flexibility, agility, sustainability) [7]. Subsequently, most PM's researchers recognize the lack of consensus around PM systems' designing and applicability [5, 6, 9, 10]. Nevertheless, they adopt open and suggestive frameworks for PM's characteristics and future considerations.

Altogether, these researchers point future directions to contemplate network performance along internal structure (i.e. collaborative teams) and external environment (i.e. supply chain); dynamic and learning systems for performance management to comply with the disruptive and transformational change on business trends; and much focus on measuring the intangible and social aspects as organizations are progressively depending on knowledge and multi-department teaming for thriving on the competitive business world.

Focusing on the operational environment, Operational Performance Management Systems (OPMS) need to follow these trends in order to create effective and agile management systems which can cope with the dynamicity of the real world and help on enhancing the daily operational performance.

Traditionally, operation managers relied on report-centric performance management in which, within a regular periodicity (e.g. monthly, weekly, daily), managers analyzed the compliance of performance indicators meeting their targets for the various operation's key functions and proposed changes based on the perceived causes for the target deviation [11]. Conversely, on the more recent years, more dynamic and informative methods have been utilized to provide a faster, simpler and more helpful way of portraying the correct information for a proper operational performance management [12]. These methods comprise the use of dashboards and interactive Business Intelligence tools which take advantage of the current ICT developments to portray the information needed, at the right time, for managers to analyze it.

Additionally, in order to know the crucial information that needs to be portrayed to operation managers, there must be a deep knowledge about the specific operational environment and how it connects with the rest of the organization's efforts, so the overall business performance can be optimized. Subsequently, [13] took a design-led approach to link strategic and operational activities to enhance the dissemination of KPI's management and accountability and change the way PM is perceived and integrated within an organization.

Therefore, the diagram on **Figure 1** illustrates the three main points operation managers should focus when developing and managing an OPMS.

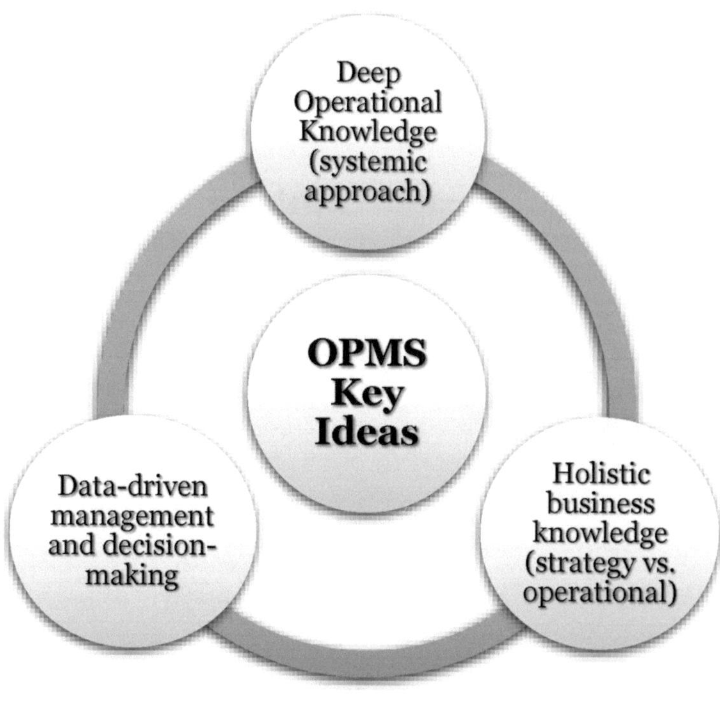

Figure 1.
OPMS key ideas to retain for its development and management.

Firstly, there is an obvious need for a deep knowledge about the specific operational environment, with a systemic approach over the intertwined processes between internal and external boundaries, and the various interactions between people, machines, and people and machines. On the other hand, there is also the necessity for connecting the strategic and operational performance objectives, with a holistic and integrated view of the whole business. Finally, the operation managers can and should leverage the ICT developments to help reduce unfounded and biased decision-making and provide an automatic integration of the various performance systems to make up a systemic OPMS composed by function-specific operational performance management sub-systems.

2.2 Design thinking and data analytics for developing OPMS

To approach these OPMS issues, the authors explored two different theoretical backgrounds which could be combined to provide an adequate solution for the prior OPMS issues, at the same time they could be leveraged to overcome each theory's intrinsic limitations. As such, the author initially summarizes each theory's advantages and limitations on **Figure 2**.

This diagram depicts both DT and DA positive and negative points and it can be noticed there are two divergent ways of analyzing the real world according to these methodologies. To the left, Design Thinking takes a broad approach to problem-solving, where it explores different problem-solution frames, through intensive validation and by gathering a vast business knowledge allied with a user-centric approach.

As limitations, some argue over the designer's intrinsic subjectivity on choosing the procedure's iterations and artifacts used, which affects the traditional business' requirement for milestones' definition and compliance. Furthermore, there is a fundamental difference on design thinking and manager thinking methodologies for problem-solving, which can sometimes lead to misunderstandings and attrition on the organizational sphere.

On the other hand, Data Analytics is more focused on analyzing a specific perceived problem or system, with great power for creating valuable and meaningful

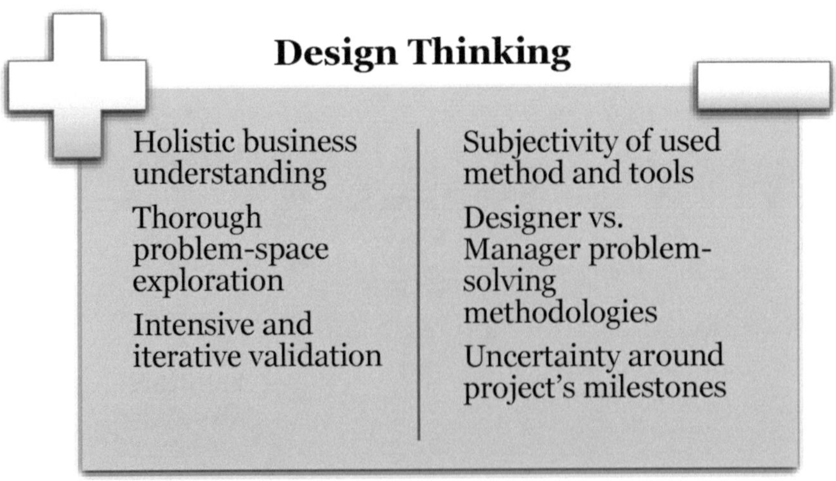

Figure 2.
Pros and cons of design thinking and data analytics methodologies.

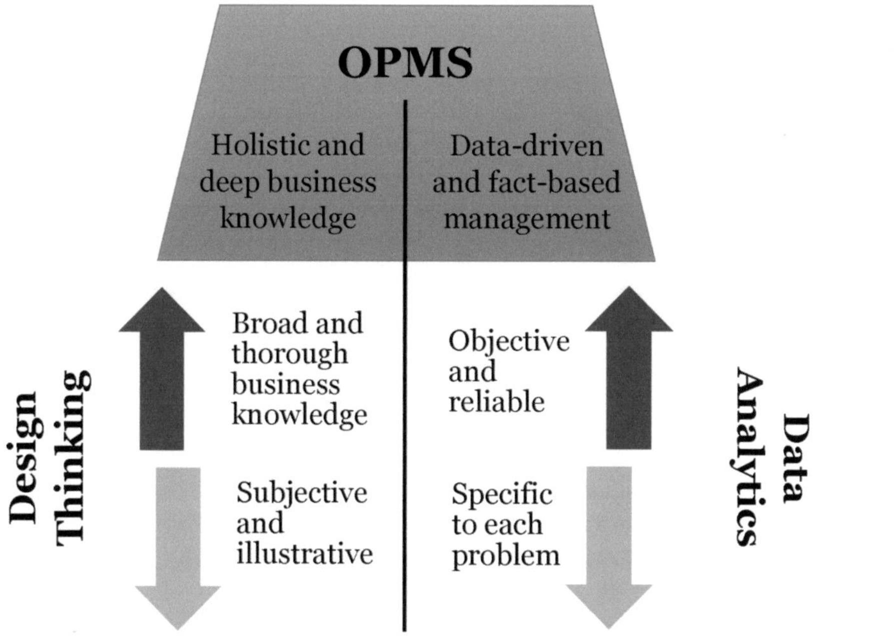

Figure 3.
Overall representation of DT and DA leveraging points for fulfilling OPMS requirements.

information for supporting decision-making. It uncovers trends and patterns through immense computing power and algorithmic data processing which could not be accomplished with mere human capabilities.

As limitations, there is a fundamental need for exploring the problem-space to prevent tacking the wrong problem and to lead to the development of a faithful and robust model. Moreover, this type of analysis is very influenced by the intrinsic data quality and the reliable representation of the reality through the captured data. Finally, it needs a dedicated data pipeline to capture, clean and structure the collected data so that different analytic tools can be explored and tested on a flexible and effortless way.

Concluding, the authors propose a combination of these two methodologies to allow for a complete and reliable way of improving the operational performance management systems. The relations and leveraging points between both methodologies and OPMS needs are portrayed on **Figure 3**. Thus, it illustrates that DT is intended to provide the holistic and deep business knowledge, while DA is supposed to provide a data-driven and fact-based management. Additionally, their limitations are conversely minimized by the complementary methodology, making this a win-win methodology.

3. Methodology proposal

After exploring the perceived topics' strengths and limitations and presenting the rationale for combining DT and DA to enhance the operational performance management systems, the methodological stages are laid down on **Figure 4**.

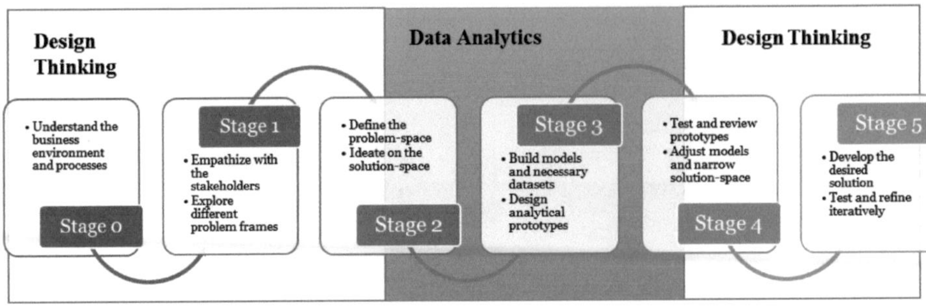

Figure 4.
Schematic representation of the methodological stages and their foundations.

This diagram portrays six main stages. The first two stages (Stages 0 and 1) are only related to Design Thinking steps. These refer to the initial mapping of the organization's activities and stakeholder's relationships, while understanding their emotional and intuitive ways of reasoning. Altogether, this allows for multiple framing of the problem-space according to different perspectives.

The third stage (Stage 2) is intended to define the problem-space as it is perceived, and explore the solution-space, by conceiving different solutions for further narrowing and prototyping. This step is presented as being included both in DT and DA methodologies since the problem definition is based on DT practices, while for the solution space exploration the practitioner should already have a DA focus to try to come up with data-driven solutions.

The fourth stage (Stage 3) is focused on the models' conception for the solutions idealized. To materialize these models, they require the development of proper datasets, along with and its inherent collection and storage processes. For further analysis and construction of the virtual model of the real system, this stage is primarily based on DA practices and tools, and it is a laborious and time-consuming step, since it needs diverse knowledge and infrastructure to capture, pre-process, structure and test different datasets for various analytic tools.

The fifth stage (Stage 4) requires a combination of DA and DT techniques and expertise to test and absorb feedback, so the developed prototypes and models can be improved or selected, grounded on a collaborative creation. This step's outcome is intended to filter misconceptions and align stakeholder's expectations, as well as narrowing down the solution space to concentrate the team's efforts and walk towards the final solution.

The sixth and final stage (Stage 5) proposes the development of the desired solution into a real-life testable model. For this purpose, it is recommended the iterative testing and refining of the perceived solution until all parts are satisfied with the product/service.

This methodology entails an inherent iterative nature between all of the six stages, repeatedly exploring and refining the problem and solution space until the team is satisfied with the frames developed.

Beyond the methodological procedure, there must be a set of tools or practices which guide the practitioner, or at least are used as reference, for applying the conceptualized methodological steps. The proposed tools and practices presented on **Table 1** are extracted from prior literature research, but these are not exclusively limited to its assigned stage, nor are they to use strictly as the literature proposes.

Stage	Tools and practices	Reference literature
Stage 0—Constructing the organizational 'picture'	Activity system mapping	[14–16]
	Competitor activity system mapping	[14, 15];
	Stakeholders mapping	[14, 17, 18]
	Value Chain analysis	[19]
	PEST analysis	[14, 20];
	Five Porter's Forces	[21]
Stage 1—Exploring users' pains and needs	Shadowing	[14, 22, 23]
	Semi-structured interviews	[14, 24]
	Customer narratives/ storytelling	[14, 24]
	Scenarios	[15, 22];
	Personas	[14, 18, 25]
	Customer journey	[14, 15];
	Mind mapping	[14]
Stage 2—Co-evolution of problem and solution space	Motivational mapping	[14, 15];
	Metaphors	[14]
	Brainstorming	[15, 26]
	Sketching	[15, 27];
	Concept development	[28, 29]
Stage 3—Solution exploration	Data types and data collection;	[30, 31];
	Data pre-processing	[30, 32];
	Data warehousing	[30, 33, 34]
	Statistical tools and models	[30, 35];
	Machine Learning models	[36, 37];
	Business analytics and visualization techniques	[38–40]
Stage 4—Prototype and test	Rapid prototyping	[15, 25, 41]
	Scenarios	[15, 22];
	Role-playing	[14, 15];
	Co-creation	[14, 28, 42]
	Future customer journey	[14, 43];
	Iterative prototyping	[14, 41];
Stage 5–Develop concrete product/ service solution	Co-creation	[14, 28, 42, 44]
	Iterative prototyping	[14, 41];
	Value delivery	[2, 14, 15, 45, 46];
	Integration with existing operations	[13, 14, 47, 48]

Table 1.
Summary of tools and practices to perform each methodological stage.

4. Case study

The proposed methodology was explored on an empirical case study of an airport's Baggage Handling System (BHS) of a European airport. This application followed the methodological procedure presented in **Figure 4** and used some of the tools and practices presented on **Table 1** to help on making sense out of the immense accounts of information gathered.

4.1 Stage 0—Constructing the organizational 'picture'

This initial stage had the objective of giving an overview of the case study environment, and it was divided in two main topics: an overall industry contextualization; and a specific case study contextualization.

The first topic started with an overview of the air transport industry and its current trends, followed by a deep-down into the airport environment. Here, a brief overview over its control and supervision was made, and some generic characteristics were laid down concerning multi-airport systems, international differences, stakeholder's mapping and economic chain, and the main regulatory drivers. Afterwards, the focus was directed to the airport's Baggage Handling System (BHS), where a generic description of its objectives, work process and primary functions was made, along with the specific regulatory drivers and upcoming regulatory frames on BHS industry.

The second topic intended to be company-specific within the case study boundaries. It started by giving a technical description of the equipment and controlling systems used for its operation. Then, the specific BHS ecosystem was portrayed, along with its main stakeholders and their influence/impact on BHS operations. Following, the BHS' operational performance and its main dimensions were addressed. Afterwards, a thorough description of the operational characteristics and the workflow along the various stakeholders involved on a baggage journey was made, depicting the intrinsic dependencies between the various BHS stakeholders. At last, an organizational overview was done showing the overall structure and the in-site hierarchy of operation and management, as depicted on **Figure 5**. Subsequently, a final description of the BHS operational management is addressed, in order to understand how the daily processes and concerns were managed and resolved.

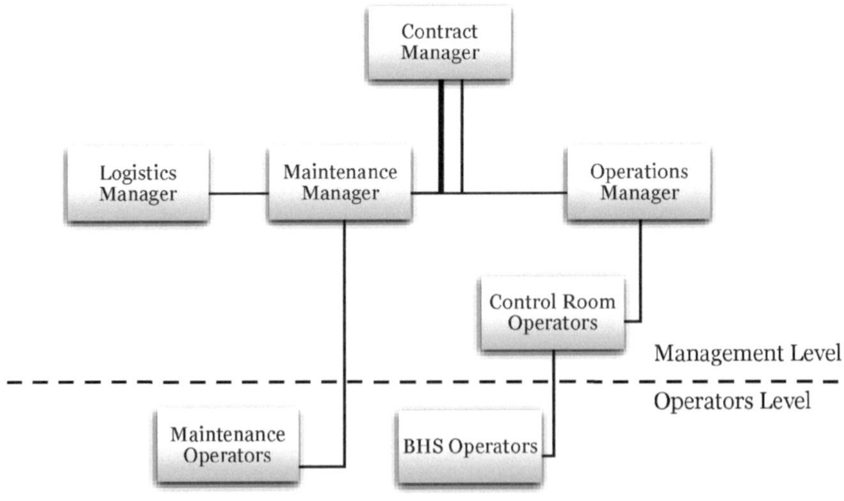

Figure 5.
In-site organizational hierarchical structure of BHS management.

4.2 Stage 1—Exploring users' pains and needs

After constructing the initial industry picture and understanding the surroundings where the BHS is included and its basic characteristics, the next step was to empathize with the various stakeholders which interact or impact the system's operations. A collection of tools was employed throughout this stage to harness on the gathered data and provide profound insights about the intuitive and relational aspects of human relations and people's reasoning when performing their jobs.

Approximately 7 hours of shadowing were undertaken to carefully observe BHS control room operators while operating the BHS. Here, the researcher had the opportunity of first-hand contact with the real-time management, which allowed him to pose some specific questions about the interactions between the CR operators and the information systems utilized for the BHS' operations management. Additionally, several semi-structured interviews with the BHS' main stakeholders were conducted to obtain a rich and broad view from the various perspectives that come into play when taking BHS operation performance into account. A total of 12 interviews were performed by the researcher, segregated according to the **Table 2**.

Each semi-structured interview was used to construct a mind map that allow for a summary of the gathered information and themes discussed, which helps to mentally process and reflect over them to generate deeper insights. Alongside these interviews, the author also performed a total of five customer narratives with different passengers, with the objective of understanding some common pains and needs that the final customer of the whole baggage journey could have. In order to get a broader picture and minimize biases throughout these collected stories, the author diversified the customers' profiles along the five customer narratives (e.g. old vs. new, frequent flyer vs. novice, family travel vs. business).

A total of seven personas were constructed in order to have a concise and coherent definition of the information gathered for further presentation and discussion with a focus group composed by experts. They comprised the main concerns and the global opinion between the interviewees belonging to each stakeholder's persona. Some stakeholder's categories were aggregated according to the relative importance to the BHS operational environment and management, resulting on the personas' set depicted on **Figure 6**.

4.3 Stage 2—Defining and ideating on the problem-solution space

This stage focuses on developing an iterative and collaborative environment for the discussion and refinement of the perceived problem space, as well as for the exploration of the solution space. This was an iterative process of exploration and discussion, in order to clarify and improve the problem definition and consequently, the solution ideation.

Internal Stakeholders						**External Stakeholders**	
Operational				**Back-office**		**Airport**	**Ground Handling**
CR Operators	Op. Manager	Contract Manager	Sales	Digital Business	Customer Service	BTO*	Operations Manager
3	2**	2**	1	1	1	1	2

**BTO–Baggage Terminal Operations.*
***One of the managers interviewed combines the function of operations and contract manager.*

Table 2.
List of semi-structured interviews, grouped by project's stakeholder.

Figure 6.
Total stakeholders personas constructed.

The authors started by reviewing the current insights gathered during the exploration phase, to look deeper into any misconnection or discover profounder insights. Hence, it was reexamined the mind maps and personas created in the prior stage, in order to develop a motivational map. This was used to understand and connect the various stakeholders' perspectives around a central concern identified throughout different interviews and, deconstruct it to find opportunities and clear the problem definition.

After trying to find connections to continue developing the problem space understanding, there were already many pains and needs identified throughout the data collection processes. Therefore, the authors had to filter them and find the most critical ones, in order to have an efficient and effective dialog with the experts' group. With this purpose, every pain and need identified until this step were listed and a multi-criteria prioritization was made. As such, the authors created a set of four criteria consisting in:

- Relative reference frequency
- Relative occurrence frequency
- Impact on BHS operations
- Feasibility

From this exercise, five main problems were prioritized, which were then taken into discussion with a focus group of experts. This group gathered both managers and engineers with experience ranging from 5 to 20 years with this type of systems, and belonging to operational and back-office teams. The meeting started by presenting a summary of the overall organizational 'picture', followed by the presentation and discussion of the five main problems prioritized earlier. As a result, it was agreed on one most desirable opportunity to be pursued, which would be the one to be explored on the solution development stage:

"Prevent the sorters' congestion"

4.4 Stage 3—Solution exploration

At this point, the pivotal opportunity had been identified and it was the one explored throughout this stage. This opportunity was related to an internal process of the BHS, thus having available automatically generated data to leverage on analytics tools, which was one of the essential research objectives. This stage had different progressive steps in order to delve deeper into this opportunity and approach different solution possibilities, according to their desirability and feasibility in terms of time and data resources available for analysis. Additionally, the BHS information systems' functioning and their integration had to be fully understood, to comprehend the type of data being captured, and how it could be combined and leveraged with the power of analytics tools.

The first step undertaken by the researchers was to further explore the identified opportunity, using the Motivational Map tool one more time. This enabled the deconstruction of the problem into some more specific issues and concerns, as well as the identification of solution paths to take into consideration. Additionally, it also allowed the narrowing down of the solution space into more concrete and defined opportunities to be further investigated. Hence, the initial opportunity was transformed into a concrete solution to be explored:

"Sorter X congestions' comprehension – causes and impacts"

Consecutively, the authors had to build a mental model which allowed him to know the dataset he would need to have in order to tackle this opportunity and develop some analytics' prototypes. This was made with the use of a metaphor with a roundabout, through which the researcher could find an analogy between the real opportunity another common context to help him on reasoning and understanding the flow of baggage through a sorter and the possible causes for congestion.

In order to gather the necessary dataset for further exploration and prototyping on the solution space, the authors had to firstly understand the underlying information systems and the type of data collected by each, and how these could be integrated into a complete coherent dataset for analysis. To help on this task, the author resorted to the Tableau Software™, an interactive visualization software for business intelligence, which allowed to visually display various combinations of variables to understand their relations and the real meaning of the virtual captions. The initial understanding of these IS, along with the mental exercise provided by the roundabout metaphor, allowed for the author to understand what types of data he would need and could have access to, in order to comprehend the sorter's congestions. Thus, these could be summarized into:

- Data about the injection lines, i.e. the amount of baggage which enters the sorter through each injection point;

- Data about the transfer chutes, i.e. the amount of baggage which leaves the sorter through each chute;

- Data about the assigned baggage destinations, i.e. the amount of baggage inside the sorter which is destined for each transfer chutes;

- Data about the sorter's congestion, i.e. combination of operational factors which indicate the existence of the incapacity of the sorter to deliver baggage into their respective chute;

The construction of the necessary dataset resulted on a combination of different sub-sets retrieved from the various IS, as illustrated on **Figure 7**.

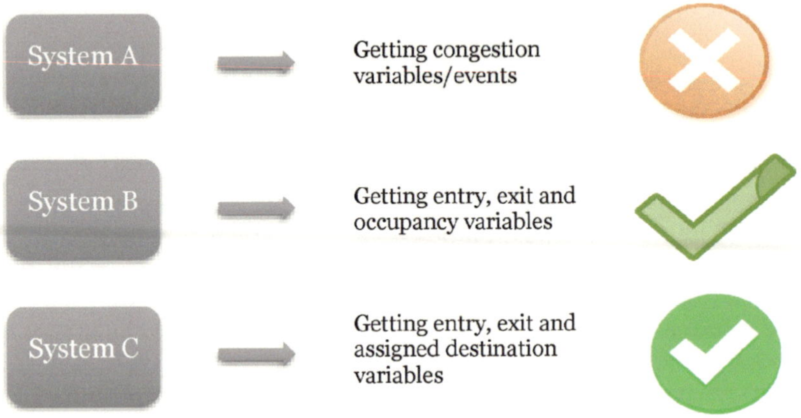

Figure 7.
Process summary to get the necessary data from the different IS.

From system A, it wasn't possible to get congestion variables since there wasn't a specific event or a coherent combination of current events being captured that could effectively indicate the congestion. Therefore, the author had to assume a different variable which could indicate congestion: the sorter's occupancy, i.e. the percentage of sorter trays being occupied relative to the total available trays. From system B, the objective was partially accomplished, since there was data characterizing the desired variables, but the time granularity with which some variables were captured wasn't enough for the detail level required. As such, the only variable possible to consider was the occupation variable. Finally, the author stepped into system C in order to collect every missing variable to complete the desired dataset. Here, the process of getting this data was more laborious and involved more complex queries in order to align the captured data for the proceeding data pre-processing. From this IS, the author could arrange the variables of the injectors, chutes and assigned destinations of the baggage going through the specified sorter.

From prior data collection, which was gathered from different IS, the author had different data types (i.e. some were counts, others were events) which could not be integrated into a whole coherent dataset to be analyzed. Furthermore, this data came with inherent redundancy, lack of normalization, and even outliers which had to be processed. Hence, different specific steps had to be taken to cleanse and integrate these loose data sub-sets into a whole dataset. Different software tools were used for the various tasks implied on this pre-processing step:

- SQL Developer™ to pre-filter and query the various databases;
- Spyder™ interactive development environment (IDE) for Python programming language to manipulate and transform data and its structure;
- Microsoft Excel™ to easily visualize and validate data and correct punctual errors;
- Tableau Prep™ to provide a visual and user-friendly manipulation for joining the various subsets and checking for redundancies;

In the end, a total of 66 variables were taken into the next stage, comprising the following variable sets:

- Injectors counts, with a total of 14 injector variables;
- Transfer chutes counts, with a total of 27 chutes variables;
- Baggage assigned destinations counts, with a total of 24 destination variables;
- Sorter's occupancy level, with one 1 variable;

4.5 Stage 4—Prototyping and testing

This stage was intended to undertake an initial analysis of the relationships within the constructed dataset, followed by the exploration of deeper analytic techniques. This procedure is necessary to build some concrete prototypes which will be tested and evaluated, thus becoming the basis for an iterative co-evolution of the final solution.

This stage unfolded in three main steps. On the first step, some descriptive statistics and the bi-variate correlations matrix were performed to better understand the variables within the dataset. This was performed using the Spearman correlation coefficient since all the variables' distributions were skewed. This is explained by the large number of zeros when there is no baggage passing through the collecting points, and the existence of a tail towards the positive integer numbers, given the relatively few but dangerous levels of high-traffic. The matrix showed some strong relations (i.e. coefficient bigger than 0.6) between some explanatory variables and the target, which is a good indicator of the possibility of building a regression model.

On the second step, the author tested some linear regressions to create an explanatory model of the sorter's functioning. From the initial dataset, encompassing every-minute counts from a one-and-a-half-month period along the 66 variables, four subsequent datasets were derived by sampling different time periods available on the initial dataset. This was aimed at mitigating the risk of overfitting, which can occur when a regression model rightfully explains the specific dataset, but it poorly performs when given a slightly different dataset. The resultant models created for each of the four datasets and their discussion gave a good overall confidence over an initial explanatory model for the sorter's functioning. Thus, considering the regression model created for the whole dataset, since it has the biggest and most complete time period, it is represented on Eq. 1.

$$\begin{aligned} \textit{Occupancy level}(\text{min}) = {} & 1{,}191 + 0{,}88 * \textit{Destination Chute}159158 + 0{,}749 \\ & * \textit{Destination Chute}151152 + 0{,}899 \\ & * \textit{Destination Chute}153160 + 1{,}008 \\ & * \textit{Destination Chute}172 + 1{,}235 \\ & * \textit{Destination MES}125 \end{aligned} \tag{1}$$

Basically, this expression provides a succinct and easy way of controlling the sorter's occupancy level and consequently, its congestions. It means that the manager can concentrate his efforts on controlling these five variables (i.e. the number baggage that is inside the sorter with destination to the respective eight transfer chutes) and, consequently, he will be able to assure the sorter does not surpass critical levels which can result on congestions. By keeping a focus on these counts and by guaranteeing the respective chutes aren't jammed, the manager can reduce the sorter's congestion probability and, consequently, the number of delayed baggage.

On the third and final step, this initial prototype was presented and discussed on a group meeting, which served as a catapult to explore the possibilities of other types of analytic models for further development, as well as to diagnose some general considerations about the overall analytic journey in order to enable future easiness and agility when trying new and different analytic ventures.

5. Results

This was the final stage conducted during this project's case study since time motives impeded its further development and conclusion into a final and feasible solution for the operational performance management. However, Stage 4 is still to be continued, according to **Figure 8**. This is an iterative stage of creating and adapting prototypes according to the various feedbacks received, with the possible return to the problem-space exploration after realizing there is missing information for each new development of the solution-space.

Afterwards, stage 5 would be responsible for pursuing a concrete and defined final solution, also by iterative development and testing with the users. These users do not comprise only the end users, but also the influencers and enablers which help on conceiving and deploying the final product/service. Furthermore, the technical, operational and financial feasibility has to be assessed and leveraged in order to create a true value-adding solution possible to be integrated and used on the operational environment, and not a solution to be used once and dropped afterwards.

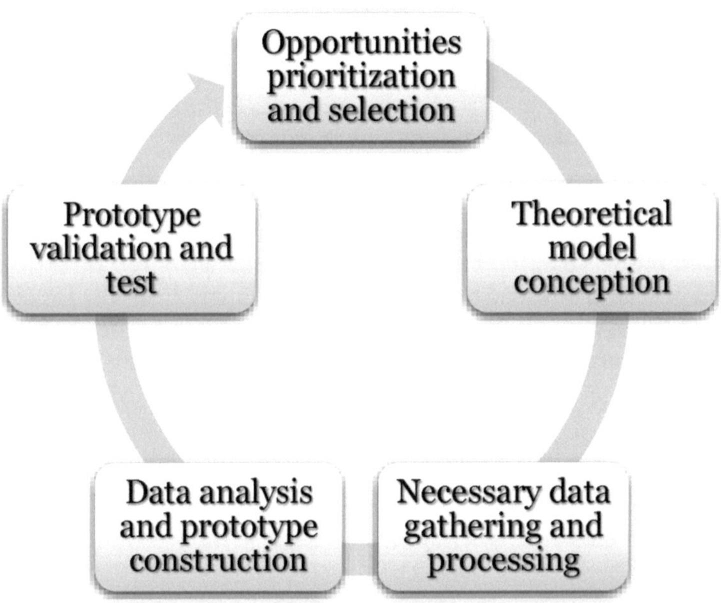

Figure 8.
Iterative process of solution exploration and testing.

6. Conclusions

This chapter proposes a hybrid Design Thinking and Data Analytics methodology for the conception of Operational Performance Management Systems (OPMS). With this purpose, an extensive literature review was made about the three topics addressed by the methodology: Operational Performance Management, Design Thinking and Data Analytics. This proposition was tested resorting to a case study research methodology, where it was possible to analyze while being deployed on a practical application. It provided evidences of the power in combining DT and DA practices for a full comprehension of the operational context, both in subjective and objective terms. Furthermore, during the empirical application, it was developed an initial prototype which conveys the capacity for readily providing operational performance improvements, as well as to envision broader possibilities which could bring huge value to the operational performance management.

Immediate operational benefits were depicted with an estimate about the savings the regression model could bring to the operational performance management, when in operation. Moreover, this prototype allowed to envision much more outcomes to the operational performance by continuing the cycle of solution exploration and refinement, with the iterative evolution of the analytic solutions and, thus, their final added value to the company. In conclusion, the proposed methodology was found successful, not only as a promising method to better understand and enhance the operational performance management, but also with many collateral effects which point towards a possible broader application. In the end, this method is perceived as completely feasible for the overall conception of holistic operational performance management systems which can take into account the various operational perspectives and needs, along with dynamic and factual tools to help on daily decision-making.

Author details

Diogo Ginjo Jantarada[1] and Antonio Grilo[2]*

1 Faculdade de Ciências e Tecnologias, Universidade NOVA de Lisboa, Caparica, Portugal

2 UNIDEMI, Faculty of Sciences and Technology, Universidade NOVA de Lisboa, Caparica, Portugal

*Address all correspondence to: acbg@fct.unl.pt

References

[1] BCG. The Most Innovative Companies in 2018. Bost. Consult. Gr., 2018. pp. 4-28. Available from: https://www.bcg.com/Images/BCG-Most-Innovative-Companies-Jan-2018_tcm9-179354.pdf [Accessed: 15 May 2018]

[2] Manyika J et al. Big Data: The Next Frontier for Innovation, Competition and Productivity. 2011. Available from: https://www.mckinsey.com/business-functions/mckinsey-digital/our-insights/big-data-the-next-frontier-for-innovation. [Accessed: 08 June 2018]

[3] Powell D. The Business Value of Design Thinking. THINK Blog [Online]. 2018. Available from: https://www.ibm.com/blogs/think/2018/03/design-thinking/ [Accessed: 14 August 2018]

[4] Christensen BYCM, Hall T, Dillon K, Duncan DS. "Customers' 'Jobs to Be Done'" no. September; 2016. pp. 2-14. Available from: https://hbr.org/2016/09/know-your-customers-jobs-to-be-done [Accessed: 14 May 2018]

[5] Kitaw D, Goshu YY. Performance measurement and its recent challenge: A literature review. International Journal of Business Performance Management. 2017;**18**(4):381

[6] Franco-Santos M, Lucianetti L, Bourne M. Contemporary performance measurement systems: A review of their consequences and a framework for research. Management Accounting Research. 2012;**23**(2):1-85

[7] Bititci U, Garengo P, Dörfler V, Nudurupati S. Performance measurement: Challenges for tomorrow. International Journal of Management Reviews. 2012;**14**(3):305-327

[8] Franco-Santos M et al. Towards a definition of a business performance measurement system. International Journal of Operations & Production Management. 2007;**27**(8):784-801

[9] Neely A. The evolution of performance measurement research: Developments in the last decade and a research agenda for the next. International Journal of Operations & Production Management. 2005;**25**(12):1264-1277

[10] Ferreira A, Otley D. The design and use of performance management systems: An extended framework for analysis. Management Accounting Research. 2009;**20**(4):263-282

[11] Slack N, Chambers S, Johnston R. Operations Management, 5th ed. Harlow, England: Prentice Hall-Financial Times; 2007

[12] Yigitbasioglu OM, Velcu O. A review of dashboards in performance management: Implications for design and research. International Journal of Accounting Information Systems. 2012;**13**(1):41-59

[13] Bucolo S, Wrigley C, Matthews J. Gaps in organizational leadership: Linking strategic and operational activities through design-led propositions. Design Management Journal. 2012;7(1):18-28

[14] Fraser HMA, Works D. How to Tackle your Toughest Innovation Challanges through Business Design. Toronto: Rotman-UTP Publishing; 2012

[15] IDEO. Design Thinking: A Method for Creative Problem Solving. IDEO-U, 2018 [Online]. Available: https://www.ideou.com/pages/design-thinking [Accessed: 28 August 2018]

[16] David HH, Ronrer-Murphy L. Activity theory as a framework for designing constructivist learning environments. Educational

Technology Research and Development. 1999;**47**(1):61-79

[17] Bourne L, Walker DHT. Visualising and mapping stakeholder influence. Management Decision. 2005;**43**(5):649-660

[18] Chasanidou D, Gasparini AA, Lee E. Design Thinking Methods and Tools for Innovation. Lecture Notes in Computer Science (Including Subseries Lecture Notes in Artificial Intelligence and Lecture Notes in Bioinformatics). Vol. 9186, No. August. 2015

[19] Porter ME. The Competitive Advantage: Creating and Sustaining Superior Performance. NY: Free Press; 1985

[20] Peng GCA, Nunes MB. Using PEST analysis as a tool for refining and focusing contexts for information systems research. In: ECRM 2007: 6th European Conference on Research Methodology for Business and Management Studies. 2007. pp. 229-236

[21] Porter ME. The Five Competitive Forces That Shape Strategy. 2008. Available from: https://hbr.org/2008/01/the-five-competitive-forces-that-shape-strategy [Accessed: 10 August 2018]

[22] Chamorro-Koc M, Adkins BA, Bucolo S. Using scenarios to explore the social aspects of design led innovations. In: Project Design 2012: Global Design Bridge. 2012. pp. 186-191

[23] Price R, Wrigley C. Design and a deep customer insight approach to innovation. Journal of International Consumer Marketing. 2016;**28**(2):92-105

[24] Rau C, Zbiek A, Jonas JM. Creating competitive advantage from services: A design thinking case study from the commodities industry service design thinking can provide the tools to help companies design value propositions that meet customer needs and sustain competitive advantage. Research-Technology Management. 2017;**60**(3):48-56

[25] Williams I et al. A collaborative rapid persona-building workshop. International Journal of Sociotechnology and Knowledge Development. 2014;**6**(2):17-35

[26] Seidel VP, Fixson SK. Adopting 'design thinking' in novice mulitdisciplinary teams: The application and limits of design methods and reflexive practices. Journal of Product Innovation Management. 2013;**30**(S1):19-33

[27] Kimbell L. Rethinking design thinking: Part II. Design and Culture. 2012;**4**(2):129-148

[28] Dorst K, Cross N. Creativity in the design process: co-evolution of problem-solution. Design Studies. 2001;**22**(5):425-437

[29] Scherer JO, Kloeckner AP, Ribeiro JLD, Pezzotta G, Pirola F. Product-service system (PSS) design: Using design thinking and business analytics to improve PSS design. Procedia CIRP. 2016;**47**:341-346

[30] Hu H, Wen Y, Chua TS, Li X. Toward scalable systems for big data analytics: A technology tutorial. IEEE Access. 2014;**2**:652-687

[31] Celano L. 6 methods of data collection and analysis. Monitoring, Evaluation, Accountability and Learning. 2014:1-30. Available from: https://www.open.edu/openlearncreate/mod/resource/view.php?id=52658 [Accessed: 12 August 2018]

[32] Lenzerini M, Salaria V, Roma I. Data integration: A theoretical perspective data integration - A theoretical perspective. 2002;**2014** (January). Available from: https://www.

cs.ubc.ca/~rap/teaching/534a/readings/Lenzerini-pods02.pdf [Accessed: 18 July 2018]

[33] Van Der Aalst W. Process Cubes: Slicing, Dicing, Rolling Up and Drilling Down Event Data for Process Mining. 2013

[34] Chaudhuri S, Dayal U. An overview of data warehousing and OLAP technology. ACM SIGMOD Record. 1997;**26**(1):65-74

[35] Sivarajah U et al. Critical analysis of big data challenges and analytical methods. Journal of Business Research. 2017;**70**:263-286

[36] Kotsiantis SB. Supervised Machine Learning: A Review of Classification Techniques Informatica; 2007;**31**: 249-268. Available from: https://datajobs.com/data-science-repo/Supervised-Learning-[SB-Kotsiantis].pdf [Accessed: 18 August 2018]

[37] Salian I. Difference between Supervised, Unsupervised, and amp; Reinforcement Learning. NVIDIA Blog [Online]. 2018. Available from: https://blogs.nvidia.com/blog/2018/08/02/supervised-unsupervised-learning/ [Accessed: 02 September 2018]

[38] Barceló J. Analytics and the art of modeling. International Transactions in Operational Research. 2015;**22**(3):429-471

[39] Velcu-Laitinen O, Yigitbasioglu OM. The use of dashboards in performance management: Evidence from sales managers. International Journal of Digital Accounting Research. 2012;**12**:39-58

[40] Schniederjans MJ, Schniederjans DG, Starkey CM. Business Analytics Principles, Concepts, and Applications with SAS: What, Why, and How. Pearson Education; 2015. Available from: http://ptgmedia.pearsoncmg.com/images/9780133989403/samplepages/9780133989403.pdf [Accessed: 20 July 2018]

[41] Grácio HL, Rijo C. Design thinking in the scope of strategic and collaborative design. Strategic Design Research Journal. 2017;**10**(1):30-35

[42] Plattner H, Meinel C, Leifer L. Design thinking research: Studying co-creation in practice. 2012;**2**(2):1-277. DOI: 10.1007/978-3-642-21643-5

[43] Vianna M, Adler IK, Vianna Y, Lucena B, Russo B. Design Thinking Design Thinking. Rio de Janeiro: MJV Press; 2011

[44] Gestwicki P, McNely B. A Case Study of a Five-Step Design Thinking Process in Educational Museum Game Design. Meaningful Play Conf. Proc; 2012. pp. 1-30. Available from: https://www.semanticscholar.org/paper/A-case-study-of-a-five-step-design-thinking-process-Gestwicki-McNely/868d5c08904213b44e0b52af9117e7e27f2d2c1b [Accessed: 29 July 2018]

[45] Geissdoerfer M, Bocken NMP, Hultink EJ. Design thinking to enhance the sustainable business modelling process—A workshop based on a value mapping process. Journal of Cleaner Production. 2016;**135**:1218-1232

[46] Henke N et al. The Age of Analytics: Competing in a Data-Driven World. 2016. Available from: https://www.mckinsey.com/business-functions/mckinsey-analytics/our-insights/the-age-of-analytics-competing-in-a-data-driven-world [Accessed: 10 July 2018]

[47] Eckerson WW. Performance Dashboards Measuring, Monitoring, and Managing your Business. 2012. Available from: https://www.semanticscholar.org/paper/Performance-Dashboards%3A-

Measuring%2C-Monitoring%2C-and-Eckerson/955805e8221c49c8a2a4056624aaab306fcdf06c [Accessed: 10 July 2018]

[48] Price R, Wrigley C, Dreiling A, Bucolo S. Design LED innovation: Shifting from smart follower to digital strategy leader in the Australian airport sector. In: Proceedings 2013 IEEE-Tsinghua International Design Management Symposium Design Business Innovation TIDMS 2013. 2014. pp. 251-258

Chapter 8

Exploring the Project Risk Management: Highlighting the Soft Side of Project Management

Muhammad T. Hatamleh

Abstract

The majority of the approaches to managing project risk follow the logic of process groups. Project Management Institute (PMI) has 29 tools and techniques related to risk management process groups. Consequently, engineering and business schools have been accused of educating managers with sharp analytical skills but little understanding of social problems. The literature suggests that too much attention is focused on learning the techniques and formalities of risk management but not enough on the advanced issues of management. Also, the literature argues that there are two approaches to project management (hard and soft). The hard side only covers part of the managerial aspects which helps to manage foreseeable uncertainties. However, unforeseeable uncertainties need skills that related to soft side approaches such as emotional intelligence, navigating the organization's culture, risk attitude, participative leadership style, and managing the relationship with stakeholders. This study provides an intensive review of the literature to discuss the need for integrating the hard and soft sides of management to achieve an effective risk management process. In addition, it proposes a conceptual framework that provides guidelines to enhance overall risk management efficiency.

Keywords: risk management, unforeseeable uncertainties, risk attitude, emotional intelligence, organization's culture

1. Introduction

The last decade showed an industrial engineering growth toward nontraditional industries, particularly information technology (IT) and service-related industries that add considering technical, organizational, ethical, social, legal, and economic factors to the project management process [1–6]. Moreover, Industrial engineering is known for creating new systems to solve problems related to waste and inefficiency [3].

Project management is a critical area of knowledge in the Industrial engineering curriculum around the world [7]. An increasing number of private and public organizations start adopting formal principles of project management to develop and deliver new or improved products, services, and organizational process changes [8]. Further, many researchers investigate ways to develop and enhance the organization's management practice [9].

The current studies of the project management process focus strongly on project risk management [10]. The current focus toward integration between partners, lean production, and outsourcing within industrial engineering has led to increase uncertainties and spike the number of accidents in the industry [11]. Consequently, organizations are giving more attention and value to risk management for improving project efficiency and effectiveness [12]. In [13–16], risk management was described as a continuous process that supports the completion of the project on time, within budget, to the required quality, and with proper provision for the safety and environmental standards.

The relationship between risk management and project success or failure has been studied extensively in the last decades [10, 12, 14, 17–20]. However, the risk management process shows a wide gap between theory and practice [21, 22]. The theory focuses on learning the techniques, planning methods, and formalities of project management while unintentionally overlook the nontraditional soft approach of management [2, 3, 22–24].

Effective project risk management obliges a wide-ranging involvement and integration across all segments of the project teams and their environment [25]. The results of previous studies that focused on risk management impact on the project success show that there are contradictions in the findings [10]. This contradiction can be explained by the tendency of the researchers to neglect aspects of uncertainty management such as considering the soft side of project management and its impact on the overall performance [10, 22].

The project management process can be categorized into hard and soft sides/approaches/skills [22, 26–31]. The hard skills focus on applying tools and techniques within project management and usually described as a science and comprise processes [26]. On the other hand, the soft skills are largely intangible, not associated with a deliverable or a concrete output, and enable working through and with people along with handling the associated human factors [22, 30].

The distinction between the concept of risk and uncertainty is still not fully clear in the context of project management [10, 32, 33]. In [34] the uncertainty was classified according to the project management techniques related to it into variation, foreseen uncertainty, unforeseen uncertainty, and chaos. Consequently, the management approach must be generated according to the types of uncertainties [10, 32, 34, 35]. In [10], it was suggested that the hard side of risk management covers the part that can manage variation and foreseen uncertainties while unforeseen uncertainties and chaos need other skills that related to the soft side.

The mere existence of accepted principles, well-defined processes, and widespread practice is not sufficient to guarantee success [24]. The hard skills just started to be viewed by many organizations more as baseline competencies rather than an additional practice that needed to improve the management process [27]. Moreover, the hard side should be considered as a necessity for the survival of the organization but not a sufficient tool in managing project risks [32]. Consequently, effective project management needs a balance between hard and soft skills [2, 28, 30].

Effective project risk management can be achieved depending on the involvement of the project team in the management process, which required a good understanding of the team environment [16]. Hence, the last decade presented a competitive global market creating a changing work environment that demands engineers who possess soft skills [36]. Further, the current risk management practices require investing more time and effort on the soft skills in order to advance the risk management process [10, 22, 24, 28, 37, 38].

This study aims to create a deeper understanding of the project risk management process by exploring the literate to investigate the potentials of integrating the hard and soft sides of risk project management. In addition, propose a broader conceptual framework for assessing and enhancing the project risk management process by including comprehensive risk management tools and techniques adopted from both sides.

2. Methods

This review covers academic publications based on theoretical and empirical findings on the concept of Risk Management. Literature was obtained through electronic searching strategy from major databases available to the researchers. The database used for this research includes but not limited to Web of Science, Science Direct, and Scopus. In other words, the database offers extensive studies on the risk project management process. However, relatively few studies were to be found concentrating on the soft skills acclimation and integration during the risk management process.

The initial search was broad enough to allow for as many results as possible including words that could identify as part of the risk management process. English was chosen to be the medium of communication. Therefore, publications that are not written in English were excluded. Moreover, this research focuses on the current research studies, studies that were published after 2005, unless if specific research offers a unique point of view or valuable contribution. **Table 1** shows the distribution of the literature, which were included in this research, based on its sources of publications.

Finally, qualitative and quantitative methods were used to analyze the review findings. The qualitative method presents a broad narrative of the findings from the literature, while the quantitative method was used in presenting the findings with tables and figures.

Literature source	Frequency
International Journal of Project Management	9
International Conference on Industrial Engineering and Engineering Management. IEEE	4
John Wiley & Sons, Inc.	4
Journal of Loss Prevention in the Process Industries	3
Risk Management	2
Safety Science	2
Project Management Journal	2
Informing Science & Information Technology	2
Scientia Iranica	2
Engineering Management Journal	1
European Journal of Engineering Education	1
European Journal of Industrial Engineering	1
PMI Global Congress EMEA Proceedings	1
Association for Project Management (APM)	1
International Conference on Industrial Engineering Theory	1
International Journal of Industrial Engineering and Management (IJIEM)	1
Project Management Institute	1
Others	45

Table 1.
The distribution of the literature based on its source.

3. Project risk management

In [20], the risk management process was defined from the literature as a formal orderly process for systematically identifying, analyzing, and responding to risk events throughout the life of a project to obtain the optimum or acceptable degree of risk elimination or control and to achieve the project objectives. Further, project risk management is known as an integrative process, where it continues throughout the project life cycle [10, 16]. However, the intensity of the risk management process might decrease as the project progresses, but still, the threat of an unforeseen emergent risk should not be ruled out until the project is completely over [15]. Hence, this definition predominantly emphasizes the hard skills at the expense of soft skills [28].

3.1 The hard side of risk management

There is a wide consensus on the indispensable elements for a risk management process [24]. This can be observed by the growing range of proficient tools and techniques, research base, and practical implementation across many industries [16, 24]. The literature offers several risk management standards, such as Risk Management Standard by the Institute of Risk Management (IRM); Project Risk Analysis and Management (PRAM) by the Association for Project Management (APM); the Project Management Body of Knowledge (PMBoK®), Chapter 11, by Project Management Institute (PMI); and Risk Management—Principles and Guidelines by the International Organization for Standardization (ISO) [16, 39–41]. These standards have well-defined processes and widespread practices that originally cover the hard side of management with few exceptions. For instance, PRAM identifies the functional roles of the organization's members during the risk planning process and considers it as an element of risk management.

The hard side of risk management demonstrates pre-specified approaches that have tools and techniques within four major process groups (identifying, analyzing, developing a response, and monitoring and controlling risks) and it is feasible if adequate information were available [10, 32, 39]. In the last decades, these process groups branched out from the original four groups into identification, qualitative analysis of risks, quantitative analysis of risk, risk response planning, and risk monitoring and control [12, 16].

In the early stages of the project, risk identification should be implemented, and any further process in risk management would be performed on these identified potential risks [42]. Therefore, all possible sources of potential risks must be identified as early as possible to help the organizations in choosing the suitable strategy [11]. One of the best methods to identify the risks is by developing a checklist categorizing the risks that might evolve during the project [18]. Also, historical records (lessons learned) and knowledge of risks from the experience-based of project personnel should be gathered and reviewed [12, 24, 43]. Further, the risks should be identified and classified by its nature and its potential impact on the current projects [16].

For risk identification, it seems that there is an agreement between researchers on the meta-classification approach, which identifies the risk factor based on three levels according to the project lifecycle environment [11, 18, 43–45]. First, the macro-level which consists of risks sourced externally (exogenously); second, the meso level which consists of risks sourced endogenously (self-developed) and project-related; and third, the micro-level which consists of risks found in the stakeholder's relationships. The final step of the risk identification should be a risk category summary sheet by using the risk breakdown structure and checklist, wherein the participation of every individual in the management team would be integrated [16, 46].

The next stage of the risk management process is analyzing the risks. This stage, introduced by the Project Management Institute, includes qualitative risk analysis and quantitative risk analysis [16, 47]. As an intermediate process, it incorporates uncertainty quantitatively and qualitatively to evaluate the potential impact of risk [48]. In this stage, the risks with high probabilities, associated with a substantial impact on the project, should be focused on. Therefore, by the end of this stage, risk and uncertainty would be identified, then rating should be accomplished by forecasting the probability of occurrence and severity of the risk impact as well [45, 48, 49].

To estimate the probability, scholars note two main approaches: subjective judgment and objective analysis [48, 50]. Subjective judgment is done by using the experience and scrutiny to make a direct estimate which allows the management to use the logic, intuition, and experience or it can be driven by the means of an educated guess [16, 45, 46]. Objective analysis usually needs historical data. Sometimes this can be impractical since it is difficult to find comparable information [16]. Therefore, scalers such as one-in-ten and one-in-hundred are often used.

In the last decade, several methods and techniques have been developed to analyze risk on an industrial plant [50]. Risk analysis has three main requirements: recognize what to expect as output data, collect the available data, and then select a suitable method for the analysis [50]. There is an agreement that these methods can be categorized into two groups: qualitative and quantitative [16]. Further, it can be described as deterministic, probabilistic, and a combination of deterministic and probabilistic [50].

More than 60 methods and techniques were identified by the scholars, one of the most used methods to estimate the impact of the risks is the analytic hierarchy process (AHP) [51]. This technique breaks down the risks into small groups, constructs a hierarchical structure, compares the impact of each factor with other factors in the same group on a pairwise base, and allocates a comparison ratio to them. Hence, the same concept used between groups, and the final impact for each factor can be determined by multiplying the ratios. **Table 2** shows some of the common methods and techniques used to analyze risks. Consequently, the estimates should be clarified and improved on an ongoing basis [45].

Classification of risk analysis methods	Methods of risk analysis	Literature
Quantitative	Accident hazard analysis (AHA)	[16, 50, 52, 53]
Quantitative	Event tree analysis (ETA)	[50, 54–56]
Quantitative	Monte Carlo analysis	[16, 50]
Quantitative	Method organized systematic analysis of risk (MOSAR)	[50, 57]
Quantitative	Optimal risk assessment (ORA)	[50, 58]
Quantitative	Simple additive weighting (SAW)	[49, 59, 60]
Qualitative	Failure mode effect analysis (FMEA)	[16, 50, 61, 62]
Qualitative	Hazard and operability (HAZOP)	[50, 63, 64]
Qualitative	Plant level safety analysis (PLSA)	[50, 65]
Qualitative and quantitative	Technique for order preference by similarity to ideal solution (TOPSIS)	[49, 66]
Qualitative and quantitative	Complex proportional assessment (COPRAS)	[49, 67]

Table 2.
Risk analysis methodologies.

Quantifying the qualitative analysis can be performed in several ways. One of them is by integrating a specific qualitative method with the Fuzzy Analytic Process [47, 55, 61, 68]. In [47] the integration was illustrated by combining the traditional AHP with fuzzy logic by giving a fuzzy scale to the AHP crisp values. Following the fuzzy ranking technique, the fuzzy scales were converted to crisp numbers by considering α-cut and expert opinions to ensure the precision of the paired comparison, which lead to have criteria weight. By using an interval scale, a fuzzy decision is initiated to develop a matrix that would help in evaluating the risk, ranking the risks, and facilitating decision making. Typically, risk scales have a mapping matrix commonly used during the qualitative analysis [47, 51]. Further, there are five types of risk scales: nominal scales, interval scales, ordinal scales, calibrated ordinal scales, and ratio scales [51]. For instance, a nominal scale would identify the cost, schedule, and quality impact of the risk, assuming it occurs. Then, the dollar cost to remedy the problem(s) would be estimated. Finally, the product of probability and consequence (the cost to remedy) would quantify the risk to this particular project.

After analyzing the risk, risk response planning should be implemented. Hence, risk response planning was identified in [16] as "the process of developing options and determining actions to enhance the opportunities and reduce threats to the project objectives." The level of risk impact is directly related to the effectiveness of the risk response process [16, 51, 69, 70]. However, the risk response process is rarely addressed in the current research related to risk management [69].

There is an agreement between scholars that the risk response process has four strategies [16, 51]. These strategies include avoidance, transfer, acceptance, and mitigation. In addition, contingency planning could be considered as a fifth strategy [70]. Moreover, it is also considered as part of the risk acceptance strategy [16]. During the risk response process, transfer and mitigation are the only strategies that involve a real investment and require budget allocation. Consequently, proactively defining an appropriate strategy would help to improve the project outcome and may result in obtaining additional benefits [51, 70].

As mentioned earlier, the project may evolve, the risks may change, the likelihood and severity of identified risks may change, new risks may emerge, identified risks may disappear, residual risks may arise, and new risks may emerge [13, 15, 16, 45, 51, 70]. Monitoring and controlling process include: tracking the identified risks, monitoring residual risks, identifying new risks, ensuring and assessing the effectiveness of the selected risk response strategies [15, 16, 51]. Therefore, the risk monitoring and controlling process are crucial for the risk management plan, and it should be developed proactively and continually during the project life cycle [16, 45, 51].

The hard side of project management is well documented between the scholars [16, 28, 39, 41, 45, 51]. In this study, several tools and techniques were investigated. In addition, this chapter would collect the most common and efficient tools and techniques to create a framework that would help to assess the risk management process and provide a guideline to ensure an effective risk management process.

3.2 The soft side of risk management

The soft side of risk management embraces the process of managing and working with people, guaranteeing customer satisfaction with the purpose of retaining them, forming a favorable atmosphere for the project team to deliver high-quality products [31]. Further, creating a favorable atmosphere in the workplace would encourage the project team to deliver a high-quality product on time and within budget [26, 27, 30, 31]. The soft side of management aims to deliver such an atmosphere [9, 10, 31].

Several soft skills dimensions were discussed and identified by scholars for the management process [10, 22, 26, 30]. These soft skills include, but not limited to, communication skills, team-building skills, flexibility and creativity skills, leadership skills, the ability to manage stress and conflict, risk attitude, awareness of emotional intelligence, and navigating the organization's culture [9, 10, 22, 26, 27, 30, 31].

In [10], the soft approach of risk management was categorized into context, strategic approach to risks and uncertainties; risk communication and information; attitude, assignment, and relationship with stakeholders; and crisis management. However, one of the most significant success factors for an effective risk management process is the one most often lacking, an appropriate and mature risk attitude [24, 28, 71]. Both researchers and practitioners agree that the attitude of individuals and organizations has a significant impact that influences whether the risk management process would deliver what it promises [24, 71]. Consequently, it is important to not ignore the fact, that risk management is undertaken by people, acting individually and in various groups [28, 71].

Attitude refers to what motivates the decision-maker to choose responses to different situations [72]. Furthermore, attitudes often might be deeply rooted and represent the core values of individuals or groups. However, the attitude represents choices that differ from personal characteristics (they are situational responses rather than natural preferences or traits) [28, 71, 72]. Risk attitude was defined as "chosen response to uncertainty that matters, influenced by perception" [71]. Therefore, risk attitude may differ depending on a range of different influences. These influences can be identified and understood, which introduce the possibility of managing them and modify the risk attitude [71–73].

An agreement between scholars can be observed, risk attitude exists on a spectrum [24]. The response to uncertainty has two dimensions: comfort level that is divided into risk-tolerant, risk-seeking, and risk addicted; and discomfort level that is divided into risk-tolerant, risk-averse, and risk paranoid [24, 71]. Hence, different risk attitudes would trigger different responses to the same situation, since attitude drives behavior.

Risk attitudes are usually adopted sub-consciously [24]. Several practitioners are accustomed to their risk attitude to the point where they behave as if there is no choice [73]. For instance, if they consider themselves with a risk-seeking or a risk-averse attitude, they would act accordingly without assessing the current situation. On the other hand, some organizations have learned to assess each situation internally, and then choose a risk attitude which is most appropriate to the current situation to offer the best chance of reaching the project objectives [71]. Consequently, risk attitude can be integrated with the risk response process group to ensure effective risk response planning.

In [71], a process that applies emotional literacy to assess risk attitude was proposed and can be used to modify the organization's risk attitude when it is needed. Accordingly, emotional literacy is the process of using emotional intelligence components (recognize, understand, and appropriately express emotions) to manage the individual and group emotions to help them succeed.

The first step in assessing the risk attitude of an organization is assessing the individuals' risk attitude toward a situation. The proximity toward risk and the influence that an individual has can be used as a proxy measure to assess the individual influence on the organization's risk attitude [71, 74]. The literature provides several methods for stakeholder mapping that includes these two variables. For instance, in [74], the stakeholder cube method was discussed as a subjective assessment of the influence and interest of an individual and how it can affect their decision-making process.

The same concept can be used to assess the individual potential influence in a group. For example, an individual with high power (power can be gained through referent power, expert power, reward power, coercive power, and legitimate power [74, 75]) have a higher influence on the behavior and outcomes of a group [76]. At the same time, the proximity to a situation drives the individual to be more active and interested in the outcomes, which encourages to influence the organization's attitude toward a situation.

The group risk attitude is influenced by other factors than the individual's risk attitude. The organization, as a group, behavior can be influenced by group dynamics, organizational culture, national culture, and societal norms [71, 77]. The group dynamic and organizational culture can have a huge impact on the organization's risk attitude and it can lead to adopting different perspectives or risk attitudes by the group from that taken by individual members. Comparing to the individual attitude, the group attitude could be influenced to become "risky shift" where the group tends to be more risk-seeking than its individuals or "cautious shift" where the group becomes more risk-averse [74].

In addition, subconscious and unmanaged risk attitudes pose a significant threat to the ability of individuals and groups to achieve their objectives [71, 76]. Therefore, understanding how the risk attitude influences the organizational behavior and the decision-making process; being able to adopt a suitable risk attitude for each situation; and if needed being able to modify the current risk attitude, are steps that help the organization to improve their risk management efficiency [24].

The organization culture could be influence by the leadership style of the top management [74, 77]. In the last decade, several studies emphasized the importance of internal communication within the organization, where the voice of lower-level employees can offer an important source of information to organizational learning and change [73, 77–80]. Further, locally held knowledge can help in risk identification and evaluation. Therefore, top management should provide a safe environment (one that shows interest and willingness to act on the provided information), even sometime, with an incentive to encourage the employee to speak up about organizational issues and potential risks [77, 80].

In general, employees tend to be intimidated to speak up since risks tend to have negative implications and often implies a need for a change [78, 79]. In [77], it was concluded that top-management support and its openness to ideas are one of the most important circumstantial factors for the employees' inclination to provide input on potential threats and opportunities. Furthermore, a participative leadership style significantly enhances the risk management process and introduce a positive interaction and advantageous atmosphere in the workplace [73, 77, 80].

Moreover, developing a positive relationship with the project stakeholders is fundamental in the risk management process [80]. This relationship may not always protect the organization from every risk, but it can be seen as a "reservoir of goodwill" as the stakeholders place their confidence in the management team, which would help to deal with risks more effectively and ultimately contribute to the achievement of organizational goals [80, 81]. For the most part, project management literature suggests that various stakeholders, which may include individuals and organizations, may be directly or indirectly involved in the process of managing risk [80].

Kutsch and Hall [81] offer an overview of management team behaviors that tended to prevent required actions or pause any changes on the original plan, extracted from the project uncertainty management and expected utility theory (EUT), regarding the relationship with stakeholder and the leadership style of top management. These behaviors were called intervening conditions that driven from a lack of knowledge, distrust, or discomfort [82]. In addition, they can be categorized in the context of uncertainty into denial, avoidance, delay, and ignorance of uncertainty [81, 82]. Hence, these behaviors are unconscious behaviors rooted in approaches driven by the management due to fear of revealing bad news or the tendency to obey the original plan and follow procedures [81].

The causes of these actions can be traced to the perception of management on the stakeholder reaction to the information. For instance, denial of uncertainty refers to the management refusal to reveal risk-related information (that may hold negative or discomforting connotations) to other stakeholders [81–83]. Denial of uncertainty can be adopted to not expose the stakeholders to something perceived as negative which might endanger the long-term relationships with them.

Avoidance of uncertainty refers to the lack of attention to risk-related information due to insufficient trust or belief in the efficacy of that information [81, 82]. Therefore, management tends to avoid uncertainty out of fears of conflicting confidence levels about risk estimates between stakeholders. On the other hand, delay of uncertainty refers to the failure to consider or resolve risk due to lack of interest or poor general approach to project management [81, 82]. In this case, management tends to delay dealing with uncertainty to accommodate the different expectations of stakeholders about how to manage risk. Finally, ignorance of uncertainty refers to the complete lack of awareness of risk-related information by the majority of stakeholders [81–83]. This behavior can be traced back to the unwillingness to spend more resources on the scanning of the environment or the inability to scan and interpret the environment because of certain factors such as complexity and dynamics of a project [34, 81].

Ignorance and denial of uncertainty could be forestalled by increasing the tolerance of ambiguity, the experience of the management team, and the amount of control that a project manager has over internal and external factors [82]. Tolerance of ambiguity was defined in [83], p. 2, as "the tendency to perceive ambiguous situations as desirable" which refers to the extent to which an individual seeks clarity and specifies vague and unclear information to use it to improve their risk management proficiencies [82]. Several studies suggested that spending more time during the environmental analysis process for the purpose of uncertainty reduction could lead to a higher degree of tolerance toward ambiguity [34, 81–83]. Furthermore, top management with greater experience (greater accumulation of relevant historic data) may help to avoid the problem of complete unawareness of threats [82].

In addition, delay, avoidance, and denial of uncertainty may be decreased with increased project manager control over internal and external factors that affect the project [81]. Hence, if managers perceive their environment as more controllable they tend to be more proactive [82]. On the other hand, only focusing on the statistical probability of threats and their impact while ignoring any other information can be considered irrational. Therefore, top management should be prepared and willing to react to any unpredicted disruptions in the project while keeping transparency with the relevant stakeholders [34].

The impact of the intervening conditions can be beyond the control of the top management or might be initiated by a supplier or a customer or even as a result of the managers' behavior [81–83]. The top management should recognize that a rational decision-making process is required, and concealing information or ignoring uncomfortable risks is not rational and might jeopardize the long-term relationship with stakeholders [81].

This section identified several soft skills approaches and highlighted the scholars' perception of the soft side of risk management. The next section would propose and suggest practices, tools, and techniques related to the soft side to help to generate a framework that assesses the risk management process and provide a guideline to ensure an effective risk management process.

4. Discussion

In the 2000s, the literature extensively studied the hard side approaches of risk management as the main approach to managing risk, while each element of the soft side approaches was studied separately [10, 24]. This study investigates these approaches to propose a conceptual framework in order to assess the risk management process implementation and provide a guideline to improve the process by integrating practices and processes from both sides of management.

A major focus of this review is to unpack the current understanding of the soft side of risk management. Also, to investigate the benefits of adopting soft approaches in parallel with the hard ones. However, this can be problematic, given a limited study of the integration concept and its ambiguity in existing literature [12].

Few studies were to be found investigating the influence of both approaches of risk management together (soft and hard) [10]. It is true that without a proper theoretical understanding of the concept of project management soft skills, the practicality would be underdeveloped and might result in improper resource distribution. In addition, risk perceptions manly steer decisions about the acceptability of risks and the core influence on behaviors [73]. However, neither perceptions of nor attitudes toward risk could be taken as equivalents of actual behavior. Consequently, the need to integrate the soft and hard skills was recognized.

The literature provides well-proven models and frameworks to describe and assess the various dimensions of the soft side of risk (e.g. risk attitude, human factors, and emotion) separately [24]. However, these dimensions interact in powerful ways and these interactions can have a significant influence in determining the effectiveness of each separate dimension [71]. On the other hand, the hard skills considered by scholars as baseline competencies that cover only part of the managerial aspects of project uncertainties. [27]. Hence, the hard side is considered essential for the day to day operation, rather than a sufficient tool by itself for managing risks, especially, if the organization commence the risk management process through its people, acting individually and in various groups. In addition, since most projects can present unforeseeable uncertainty, this study suggests the need to integrate the hard and soft sides during the risk management process. **Figure 1** shows a conceptual framework that integrates practices and tools from both sides of risk management to ensure a more efficient risk management process.

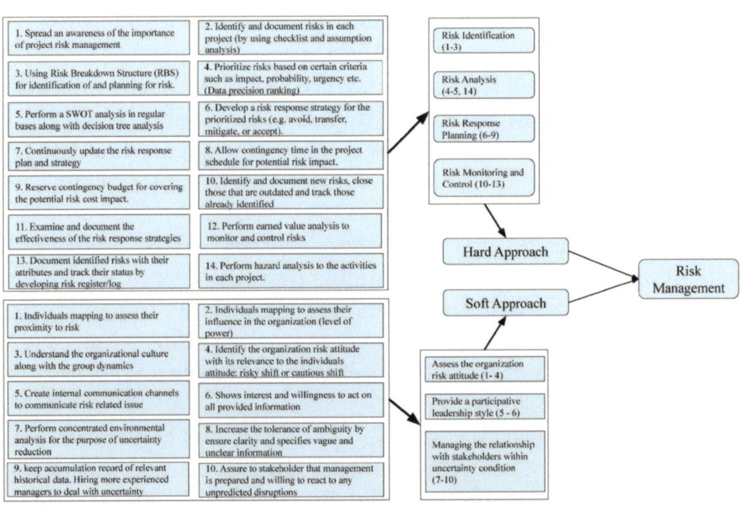

Figure 1.
Conceptual framework (comprehensive integrated tools and practices within the risk management process).

At the same time, researchers found that there is a strong relationship between the amount of risk management implemented in a project and the level of project success [10]. The earlier risk management was undertaken in a project, the higher is the level of success [17]. Consequently, organizations are giving more attention and value to risk management for improving project efficiency and effectiveness [12].

In [10], the impact of the soft approach was investigated with the relationship to project success, and they found that the soft side of risk management appears most prominently and explains 10.7% of the effect on project success. In addition, they investigate the relation to the hard side, and they found a significant correlation between the two sides. Further, they found that the soft side supports the hard side with a correlation explains 25.3% of the effect on the hard side.

The hard side has been consolidated over time with the effort of professional associations, companies, and scholars through toolsets, standards, and BoKs [10, 16, 39]. This study combined tools and practices from the literature in order to gather 14 practices and tools that best explains and covers the hard side of risk management, and it was driven from the main 4 process groups of risk management. However, scholars suggest that the impact of this effort on the project success is still weak and can be improved by integrating the soft side of management within the risk management process [10]. Consequently, as illustrated in **Figure 1**, this study focused on three aspects of the soft side to enhance the risk management proficiencies and resulted in three process groups driven from the following literature [10, 22, 24, 26–31, 71, 73, 74, 76–83]. These groups include:

- The organization risk attitude, and have four tools and practices: perform individuals mapping to assess their proximity to risk; perform individuals mapping to assess their influence in the organization (level of power); investigate and understand the organizational culture along with the group dynamics and its influence on the organization risk attitude; identify the organization risk attitude with its relevance to the individuals' attitude (risky shift or cautious shift).

- Participative leadership style, and have two tools and practices: create internal communication channels to communicate risk-related issues; show interest and willingness to act on all provided information.

- The relationship with stakeholders within uncertainty condition, and have four tools and practices: perform concentrated environmental analysis for the purpose of uncertainty; increase the tolerance of ambiguity by ensure clarity and specifies vague and unclear information; keep accumulation records of relevant historic data along with hiring more experienced managers to deal with uncertainty; assure to the stakeholders that top management are prepared and willing to react to any unpredicted disruptions in the project.

This integrated framework provides base guidelines to enhance overall risk management efficiency. For instance, using the practices from the organization risk attitude process group can help to assess, describe, and understand the organization's risk attitudes. Consequently, the action is required to modify attitude, especially, when identified risk attitude is not beneficial to achieve effective risk management [24, 71]. Further, recent studies in the field of emotional intelligence provide means that can promote and manage attitudinal change for both individuals and organizations [71].

5. Conclusion

In the last decades, scholars argued for the need of combining the hard and soft skills of management [22]. Consequently, tools and practices that encourage the hard side of management are necessary, but they need to be supplemented with leadership and soft skills. Since effective project risk management requires broad involvement and collaboration across all segments of the project team and its environment [10]. A deeper understanding of the soft side of risk management can open up a wide range of opportunities for scholars and practitioners interested in improving the risk process around the world. Therefore, the conceptual framework presented in this study provides a guide to facilitate integrating the hard and soft sides of risk management.

The majority of the literature support that the soft side has an impact on project success. Further, a significant relationship between the hard and soft sides is recognized in several fields. This relation influences the implementation of the risk management process throughout the project life-cycle. However, even that several studies consider the soft skills as requisites for success, some still disagree with the fact that these skills can be taught, learned, or managed and advocate that these skills are innate or genetic [28]. This study provides a way to assess and describe some of these skills and provides tools and techniques to influence and manage them.

Finally, focusing on the suggested three process groups helps scholars and practitioners to better understand the soft side concept of risk management and pave the way for improving the risk management process itself. In conclusion, the soft side of risk management is a viable concept within risk and uncertainty management studies, which is yet to be fully explored. In addition, integrating the soft and hard skills offer a broader risk management process that ensures more efficient results. Consequently, there is a need for more in-depth research that goes beyond documentation of meanings and activities regarding the soft side of risk management to the documentation of the process that integrates the hard and soft sides and monitors the progress resulted in implementing the integration.

Author details

Muhammad T. Hatamleh
The University of Alabama, Tuscaloosa, USA

*Address all correspondence to: mthatamleh@crimson.ua.edu

References

[1] Paté-Cornell ME. Quantitative safety goals for risk management of industrial facilities. Structural Safety. 1994;**13**(3):145-157

[2] Rabelo L. 2006-2655: Emerging topics for industrial engineering curriculum. American Society for Engineering Education. 2006;**532**(11):1-11

[3] Eskandari H, Sala-Diakanda S, Furterer S, Rabelo L, Crumpton-Young L, Williams K. Enhancing the undergraduate industrial engineering curriculum. Education and Training. 2007;**49**(1):45-55

[4] Marley RJ, Stanley LM, Muthumani A. Recent evolutions in the curricula of leading industrial engineering programs within the United States. In: International Conference on Industrial Engineering Theory, Applications and Practice; 7-10 September 2008; Las Vegas, Nevada. 2008

[5] Cooke-Davies TJ, Crawford LH, Lechler TG. Project management systems: Moving project management from an operational to a strategic discipline. Project Management Journal. 2009;**40**(1):110-123

[6] Spalek S. Improving industrial engineering performance through a successful project management office. Inzinerine Ekonomika-Engineering Economics. 2013;**24**(2):88-98

[7] Lima RM, Mesquita D, Amorim M, Jonker G, Flores MA. An analysis of knowledge areas in industrial engineering and management curriculum. International Journal of Industrial Engineering and Management (IJIEM). 2012;**3**(2):75-82

[8] Cleland DI. Project Management: Strategic Design and Implementation. 4th ed. NY, USA: McGraw-Hill; 2012

[9] Cooke-Davies T. The Wiley guide to managing projects. In: Peter W, Jeffrey K, editors. Project Management Maturity Models. NY, USA: Wiley & Sons, Inc; 2004. pp. 1234-1255. DOI: 10.1002/9780470172391.ch49

[10] Carvalho MM, Rabechini JR. Impact of risk management on project performance: The importance of soft skills. International Journal of Production Research. 2015;**53**(2):321-340

[11] Faisal MN, Banwet DK, Shankar R. Quantification of risk mitigation environment of supply chains using graph theory and matrix methods. European Journal of Industrial Engineering. 2007;**1**(1):22-39

[12] Raz T, Hillson D. A comparative review of risk management standards. Risk Management. 2005;**7**(4):53-66. Available from: http://www.jstor.com/stable/3867797

[13] Chapman C, Ward S. Project Risk Management: Processes, Techniques, and Insights. Chichester, UK: John Wiley; 1996. p. 344

[14] Hwang BG, Zhao X, Toh LP. Risk management in small construction projects in Singapore: Status, barriers, and impact. International Journal of Project Management. 2014;**32**(1):116-124

[15] Merna T, Al-Thani FF. Corporate Risk Management. NJ, USA: John Wiley & Sons; 2011

[16] Project Management Institute. A Guide to the Project Management Body of Knowledge (PMBOK® Guide). 6th ed. Newtown Square, PA, USA: Project Management Institute, Inc.; 2017

[17] Raz T, Michael E. Use and benefits of tools for project risk management. International Journal of Project Management. 2001;**19**(1):9-17

[18] Bing L, Akintoye A, Edwards PJ, Hardcastle C. The allocation of risk in PPP/PFI construction projects in the UK. International Journal of Project Management. 2005;**23**(1):25-35

[19] De Bakker K, Boonstra A, Wortmann H. Does risk management contribute to IT project success? A meta-analysis of empirical evidence. International Journal of Project Management. 2010;**28**(5):493-503

[20] Banaitiene N, Banaitis A. Risk management in construction projects. In: Banaitiene N, editor. Risk Management – Current Issues and Challenges. 2012. pp. 429-448

[21] Baloi D, Price AD. Modeling global risk factors affecting construction cost performance. International Journal of Project Management. 2003;**21**(4):261-269

[22] Söderlund J, Maylor H. Project management scholarship: Relevance, impact, and five integrative challenges for business and management schools. International Journal of Project Management. 2012;**30**(6):686-696

[23] Chen CK, Jiang BC, Hsu KY. An empirical study of industrial engineering and management curriculum reform in fostering students' creativity. European Journal of Engineering Education. 2005;**30**(2):191-202

[24] Hillson D, Murray-Webster R. Understanding and managing risk attitude. In: Proceedings of 7th Annual Risk Conference; November 2004; London, UK. Vol. 26. 2004

[25] Thamhain H. Managing risks in complex projects. Project Management Journal. 2013;**44**(2):20-35

[26] Sukhoo A, Barnard A, Eloff MM, Van der Poll JA, Motah M. Accommodating soft skills in software project management. Issues in Informing Science & Information Technology. 2005;**1**:2

[27] Van Ingen S. Leadership of project teams: Management is doing things right, leadership is doing the right things--Peter Drucker & Warren Bennis. Chemical Engineering. 2007;**114**(1):55-59

[28] Gillard S. Soft skills and technical expertise of effective project managers. Issues in Informing Science and Information Technology. 2009;**1**:6

[29] Azim S, Gale A, Lawlor-Wright T, Kirkham R, Khan A, Alam M. The importance of soft skills in complex projects. International Journal of Managing Projects in Business. 2010;**3**(3):387-401

[30] Marando A. Balancing Project Management Hard Skills and Soft Skills. Waltham, Massachusetts, USA: Rabb School of Continuing Studies, Division of Graduate Professional Studies at Brandeis University; 2012

[31] Ravindranath S. Soft skills in project management: A review. IUP Journal of Soft Skills. 2016;**10**(4):16

[32] Perminova O, Gustafsson M, Wikström K. Defining uncertainty in projects – A new perspective. International Journal of Project Management. 2008;**26**(1):73-79

[33] Fregonara E, Ferrando DG, Pattono S. Economic–environmental sustainability in building projects: Introducing risk and uncertainty in LCCE and LCCA. Sustainability. 2018;**10**(6):1901

[34] De Meyer AC, Loch CH, Pich MT. Managing project uncertainty: From variation to chaos. MIT Sloan Management Review. 2002;**43**(2):60

[35] Naderpour A, Majrouhi Sardroud J, Mofid M, Xenidis Y, Pour RT.

Uncertainty management in time estimation of construction projects: A systematic literature review and new model development. Scientia Iranica. 2019;**26**(2):752-778

[36] Kumar S, Hsiao JK. Engineers learn "soft skills the hard way": Planting a seed of leadership in engineering classes. Leadership and Management in Engineering. 2007;**7**(1):18-23

[37] Kimbrough RL, Componation PJ. The relationship between organizational culture and enterprise risk management. Engineering Management Journal. 2009;**21**(2):18-26

[38] Summerill C, Pollard SJ, Smith JA. The role of organizational culture and leadership in water safety plan implementation for improved risk management. The Science of the Total Environment. 2010;**408**(20):4319-4327

[39] ISO. ISO31000. Risk Management – Principles and Guidelines. Geneva, Switzerland: International Organization for Standardization; 2009. pp. 15-21

[40] AIRMIC ALARM IRM. A Risk Management Standard. London, UK: The Institute of Risk Management, The National Forum for Risk Management in the Public Sector, The Association of Insurance and Risk Managers; 2002

[41] Bartlett J. Project Risk Analysis and Management Guide. Buckinghamshir, United Kingdom: Association for Project Management (APM) Publishing Limited; 2004

[42] Hallikas J, Virolainen VM, Tuominen M. Risk analysis and assessment in network environments: A dyadic case study. International Journal of Production Economics. 2002;**78**(1):45-55

[43] Leung HM, Tummala VR, Chuah KB. A knowledge-based system for identifying potential project risks. Omega. 1998;**26**(5):623-638

[44] Jones K. Identifying and Managing Project Risks. ConstructConnect, Inc; 2017. Available from: https://www.constructconnect.com/blog/operating-insights/identifying-managing-construction-project-risks/

[45] Elkington P, Smallman C. Managing project risks: A case study from the utilities sector. International Journal of Project Management. 2002;**20**(1):49-57

[46] Al-Bahar JF, Crandall KC. Systematic risk management approach for construction projects. Journal of Construction Engineering and Management. 1990;**116**(3):533-546

[47] Li HX, Al-Hussein M, Lei Z, Ajweh Z. Risk identification and assessment of modular construction utilizing fuzzy analytic hierarchy process (AHP) and simulation. Canadian Journal of Civil Engineering. 2013;**40**(12):1184-1195

[48] Abdo H, Kaouk M, Flaus JM, Masse F. A safety/security risk analysis approach of industrial control systems: A cyber bowtie – Combining new version of attack tree with bowtie analysis. Computers & Security. 2018;**72**:175-195

[49] Tamošaitienė J, Zavadskas EK, Turskis Z. Multi-criteria risk assessment of a construction project. Procedia Computer Science. 2013;**17**:129-133

[50] Tixier J, Dusserre G, Salvi O, Gaston D. Review of 62 risk analysis methodologies of industrial plants. Journal of Loss Prevention in the Process Industries. 2002;**15**(4):291-303

[51] Kerzner H. Project Management: A Systems Approach to Planning, Scheduling, and Controlling. Hoboken, New Jersey, USA: John Wiley & Sons; 2017

[52] Ma L, Yan X, Qiao W. A quasi-Poisson approach on modeling accident

hazard index for urban road segments. Discrete Dynamics in Nature and Society. 2014;**2014**:1

[53] de la Huerga MR, Silvera VA, Turoff M. A CIA–ISM scenario approach for analyzing complex cascading effects in operational risk management. Engineering Applications of Artificial Intelligence. 2015;**46**:289-302

[54] Hong ES, Lee IM, Shin HS, Nam SW, Kong JS. Quantitative risk evaluation based on event tree analysis technique: Application to the design of shield TBM. Tunnelling and Underground Space Technology. 2009;**24**(3):269-277

[55] Khaleghi S, Givehchi S, Karimi S. Fuzzy risk assessment and categorization, based on event tree analysis (ETA) and layer of protection analysis (LOPA): Case study in gas transport system. World Applied Programming. 2013;**3**(9):417-426

[56] Andrews JD, Dunnett SJ. Event-tree analysis using binary decision diagrams. IEEE Transactions on Reliability. 2000;**49**(2):230-238

[57] Perrin L, Muñoz-Giraldo F, Dufaud O, Laurent A. Normative barriers improvement through the MADS/MOSAR methodology. Safety Science. 2012;**50**(7):1502-1512

[58] Khan FI, Iqbal A, Abbasi SA. Risk analysis of a petrochemical industry using ORA (optimal risk analysis) procedure. Process Safety Progress. 2001;**20**(2):95-110

[59] Jaberidoost M, Olfat L, Hosseini A, Kebriaeezadeh A, Abdollahi M, Alaeddini M, et al. Pharmaceutical supply chain risk assessment in Iran using analytic hierarchy process (AHP) and simple additive weighting (SAW) methods. Journal of Pharmaceutical Policy and Practice. 2015;**8**(1):1-10

[60] Cahyapratama A, Sarno R. Application of analytic hierarchy process (AHP) and simple additive weighting (SAW) methods in singer selection process. In: 2018 International Conference on Information and Communications Technology (ICOIACT); 6 March 2018. IEEE; 2018. pp. 234-239

[61] Sharma RK, Kumar D, Kumar P. Systematic failure mode effect analysis (FMEA) using fuzzy linguistic modelling. International Journal of Quality & Reliability Management. 2005;**22**(9):986-1004

[62] Wessiani NA, Yoshio F. Failure mode effect analysis and fault tree analysis as a combined methodology in risk management. IOP Conference Series: Materials Science and Engineering. 2018;**337**(1)

[63] Duknjó J, Fthenakis V, Vílchez JA, Arnaldos J. Hazard and operability (HAZOP) analysis. A literature review. Journal of Hazardous Materials. 2010;**173**(1-3):19-32

[64] Baybutt P. A critique of the Hazard and Operability (HAZOP) study. Journal of Loss Prevention in the Process Industries. 2015;**33**:52-58

[65] Toola A. Plant level safety analysis. Journal of Loss Prevention in the Process Industries. 1992;**5**(2):119-124

[66] Ozturk D, Batuk F. Technique for order preference by similarity to ideal solution (TOPSIS) for spatial decision problems. In: Proceedings ISPRS; 1 May 2011. Vol. 1. 2011

[67] Stefano NM, Casarotto Filho N, Vergara LG, da Rocha RU. COPRAS (Complex Proportional Assessment): State of the art research and its applications. IEEE Latin America Transactions. 2015;**13**(12):3899-3906

[68] Fattahi R, Khalilzadeh M. Risk evaluation using a novel hybrid

method based on FMEA, extended MULTIMOORA, and AHP methods under fuzzy environment. Safety Science. 2018;**102**:290-300

[69] Shoar S, Nazari A. An optimization framework for risk response actions selection using hybrid ACO and FTOPSIS. Scientia Iranica. Transaction E: Industrial Engineering. 2019;**26**(3):1763-1777

[70] Gonen A. Optimal risk response plan of project risk management. In: 2011 IEEE International Conference on Industrial Engineering and Engineering Management; 6 December 2011. IEEE; 2011. pp. 969-973

[71] Hillson D, Murray-Webster R. Managing risk attitude using emotional literacy. In: PMI Global Congress EMEA Proceedings. 2006

[72] Slovic P, Finucane ML, Peters E, MacGregor DG. Risk as analysis and risk as feelings: Some thoughts about affect, reason, risk, and rationality. Risk Analysis: An International Journal. 2004;**24**(2):311-322

[73] Rohrmann B. Risk perception, risk attitude, risk communication, risk management: A conceptual appraisal. In: 15th International Emergency Management Society (TIEMS) Annual Conference. 2008

[74] Hillson D, Murray-Webster MR. Managing Group Risk Attitude. Aldershot, United Kingdom: Gower Publishing Ltd.; 2012

[75] Pacagnella Júnior AC, Porto GS, Pacífico O, Salgado Júnior AP. Project stakeholder management: A case study of a Brazilian Science Park. Journal of Technology Management & Innovation. 2015;**10**(2):39-49

[76] Weber EU, Blais AR, Betz NE. A domain-specific risk-attitude scale: Measuring risk perceptions and risk behaviors. Journal of Behavioral Decision Making. 2002;**15**(4):263-290

[77] Sax J, Torp SS. Speak up! Enhancing risk performance with enterprise risk management, leadership style and employee voice. Management Decision. 2015;**53**(7):1452-1468

[78] Detert JR, Burris ER. Leadership behavior and employee voice: Is the door really open? Academy of Management Journal. 2007;**50**(4):869-884

[79] Ekaterini G. The impact of leadership styles on four variables of executives workforce. International Journal of Business and Management. 2010;**5**(6):3

[80] Rhee Y. Risk communication management: A case study on Brookhaven National Laboratory. Journal of Communication Management. 2008;**12**(3):224-242

[81] Kutsch E, Hall M. Intervening conditions on the management of project risk: Dealing with uncertainty in information technology projects. International Journal of Project Management. 2005;**23**(8):591-599

[82] Kutsch E. The effect of intervening conditions on the management of project risk. International Journal of Managing Projects in Business. 2008;**1**(4):606-610

[83] McLain DL, Kefallonitis E, Armani K. Ambiguity tolerance in organizations: Definitional clarification and perspectives on future research. Frontiers in Psychology. 2015;**6**:344

Section 4

Enhanced Data Analytics

Chapter 9

Linear Programming Optimization Techniques for Addis Ababa Public Bus Transport

Eshetie Berhan and Daniel Kitaw

Abstract

Anbessa City Bus Service Enterprise (ACBSE) is the only public enterprise that provides transport services in the city of Addis Ababa. The enterprise uses a fixed bus schedule system to serve passengers in more than 125 routes. However, the current bus assignment and scheduling system are becoming a challenge in the company's operational performances. The objective of this paper is to develop an optimum bus assignment method using linear programming (LP). After a thorough analysis of the existing bus scheduling system, the LP model is developed and used to determine the optimal number of busses for each route in four shifts. The output of the LP-model is then validated with the performances of the existing systems. The findings of the study showed that the new model reveals better performances on the operating costs, bus utilization, and trips and distance covered compared with the existing scheduling system. The bus utilization is improved by the new system and cut costs on the one hand and improves the service quality to passengers on the other hand. The authors recommended the enterprise to adopt the new bus assignment system so that busses can be assigned based on the demand distribution of passengers for each route at a given shift.

Keywords: bus scheduling, bus utilization, linear programming, optimization techniques, model validation

1. Introduction

The bus schedule is one of the operations planning process in bus transport that deals with the proper assignment of busses to routes to serve the expected passenger demand. The planning process in public transportation consists of different recurrent and complex tasks. It starts at a strategic level by collecting or forecasting the number of passengers at each transfer point, which is most of the time fully unknown and adds to the complexity of the planning process [1–3].

The decision-making process of the bus assignment is, however, a trade-off between service quality and operating costs for the bus operating companies [4]. It is because using too many busses incurs more operating costs while resulting in good service quality, whereas too few busses have the opposite effect. Based on the information collected from Anbessa City Bus Service Enterprise (ACBSE) currently, the enterprise serves more than 125 routes (as of 2019) that connect different parts of the city using 759 operational busses. The number of passengers shows high variability during each period which requires fluctuating the number of assigned busses in each route. But the enterprise uses mainly a fixed number of busses scheduled per route in its operation throughout the day. This resulted in some busses moving empty while others are being overcrowded, which subsequently results in poor performance and service quality. Moreover, the transportation service in ACBSE has many challenges such as low bus utilization, unsatisfied passengers' demand, and higher operating costs.

To address the challenges of bus assignment and scheduling problems in ACBSE, this paper first focuses to develop a demand-oriented Linear Programming (LP). Linear Programming is a well-accepted technique within the field of Operations Research, a specialty area within the broader field of Industrial Engineering. Then the LP-model is used to solve and optimally satisfy the existing passengers' demand in four operating periods in a day using 93 selected routes. For simplicity purpose, in this paper, the four operating periods are named as shifts.

2. Bus scheduling techniques

The concept of planning a minimum cost set of transporting routes to serve a group of customers is a fundamental constraint in the field of transport and logistics [5–7]. It is because, in the total cost of the product, transportation accounts for about 20% of the total costs of a product [8]. Therefore, the need for developing a better route plan that can reduce the cost of transportation is the concern of various industries in the field.

To address the above issues, the basic and well-studied routing model is the Traveling Salesman Problem (TSP), in which a salesman is to visit a set of cities and return to the city he started in [2, 9, 10]. The objective of the TSP is to minimize the total distance traveled by the salesman. Vehicle Routing Problem (VRP) is a generalization of the TSP in that the VRP consists of determining m vehicle, where a route is a tour that begins at the depot, visits a subset of the customers in a given order, and returns to the depot [9–11].

The activity of planning and designing a delivery or a pickup service to customers in the logistics and supply chain is known as a Vehicle Routing Problem [6]. The first time it was proposed by [12] under the title "Truck dispatching problem" to design the optimum routing of a fleet of gasoline delivery trucks between a bulk terminal and a large number of service stations supplied by the terminal. Often the context is that of delivering goods located at a central depot to customers who have placed orders for such goods, but the area of application of VRP is also so versatile and is used in many areas in real-world life.

"The VRP is defined by a depot, as a set of geographically scattered customers with known demands, and a set of vehicles with fixed capacity" [7, 13]. All depts must be visited just once and the total demand of a route must not exceed the total vehicle capacity. The VRP aims to minimize the overall distribution costs. In most real-life distribution contexts many side constraints complicate the VRP model. These side constraints can be time, that is the total route time and windows time within which the service must begin.

In the literature, VRP was also known as the "vehicle scheduling" (VSP) [6], or "Vehicle dispatching" or simply as the "delivery problem" [14]. It appears very frequently in real-world situations not directly related to the physical delivery of goods.

The VRP problem is a combination of the two well-known optimization problems: the Bin Packing Problem (BPP) and the Traveling Salesman Problem (TSP) [15–17]. The BPP is a problem given a finite set of numbers (the item sizes) and a constant K, specifying the capacity of the bin, what is the minimum number of bins needed? [16].

Logically, all the items have to be inside exactly in one bin and the total capacity of items in each bin has to be within the capacity limits of the bin. This is known as the best packing version of BPP. The TSP [3] is about a traveling salesman who needs to visit several cities. The salesman has to visit each city exactly once, start and end location, commonly called depot in VRP. The issue is to search the shortest tour within all the cities [17]. Connecting this to the VRP, customers can be allotted to vehicles by solving BPP and the order in which they are visited can be found by solving TSP. A VRP with a single vehicle and infinite capacity is a TSP.

VRP is a common name given to a problem in which a set of routes for a fleet of vehicles based at one or several locations called a depot, must be determined for several dispersed cities or customers [18]. The motive is to service a set of customers with a minimum-cost [16, 19]. Vehicle routes originate and terminate at a depot. It is one of the most challenging combinatorial optimization problems in distribution, and logistics [7]. Customers may be in a dispersed location and a fleet of vehicles need to serve them from a depot and return to the depot [16]. The decision here is to determine the assignment of the vehicle (s) and route (s) that a vehicle will serve them best. The commonly used illustration of the input and output of VRP is given in **Figures 1** and **2**.

Since both BPP and TSP are the so-called NP-hard problems and since VRP is a combination of the two, it is also NP-hard [12, 16, 20, 21]. Since the last decades, VRP has got much interest from many scholars. Even in recent years, with the rapid advancements of globalization and supply chain systems, VRP is becoming one of the important research topics in the fields [4, 22, 23].

Moreover, the complexity and its application importance immense literature have devoted to the study and analysis of Bus Scheduling Problem (BSP) and many optimization models have been proposed [23]. The different models developed have tried to accomplish near-optimal solutions with an acceptable amount of computational effort and time [6]. There are many extensions for the Vehicle Schedule Problem (VSP) or VRP with several requirements in the literature over the last 50 years [16, 24]. Among many others, some of the examples are the existence of one depot [18] or more than one depots [4, 16], a heterogeneous fleet with multiple vehicle types [18] the permission of variable departure times of trips, VRP with deterministic demand which is commonly called classical VRP [13, 18].

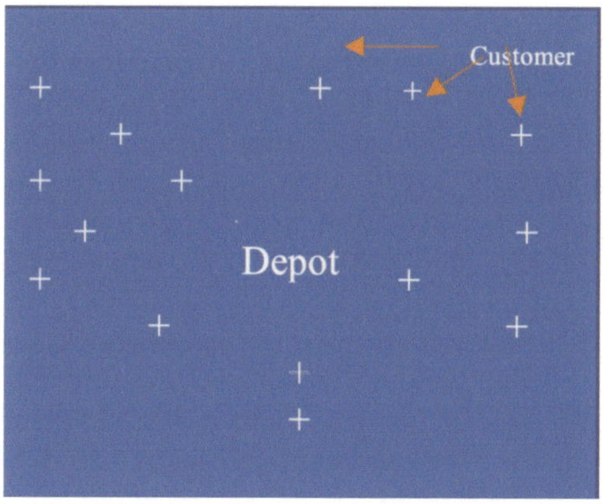

Figure 1.
VRP inputs.

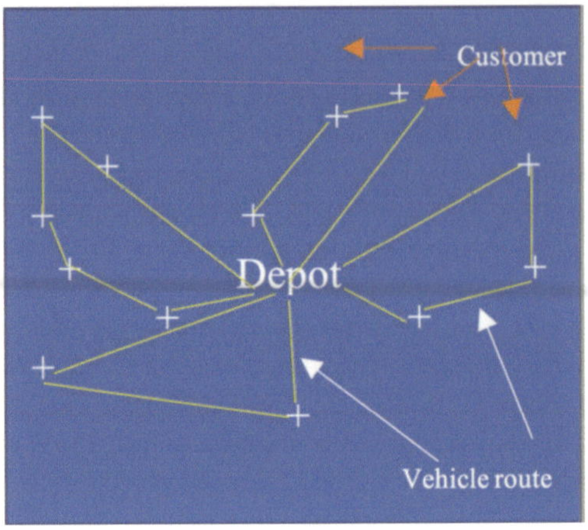

Figure 2.
VRP outputs.

3. Modeling and vehicle routing problem

Vehicle Routing Problem has been used in different applications and practices but with many constraints. The major constraints according to [16] are the network [25], demand and customers, depot locations, the type of vehicle used, and sometimes drivers. it is not possible to satisfy all the constraints in one model. In this case, some of the constraints can be reduced without loss of generality. When some customers left unserved due to this reason, in VRP, it called route failure [7]. In some cases, by introducing different penalties or priorities, the situations can be handled [6, 16].

The routing operations are performed to serve customers to start and end at one or more locations, located at the road [5]. Each location or depot is identified by the number and types of vehicles associated with it and by the number of goods it can deal with [23]. Transportation of goods is carried out by using a fleet of vehicles with fixed composition and size according to the necessities of the customers [25, 26].

The vehicle used in VRP is also one constraint in the model assumptions. According to [16], the vehicles may be characterized by the capacity of the vehicle, expressed as the maximum weight, or volume, or the number of pallets, the vehicle can carry. In most application areas, the least practice constraint is the drivers [16].

VRP is one of the most studied combinatorial optimization problems and is concerned with the optimal design of routes to be used by a fleet of vehicles to serve a set of customers [26]. VRP is not only focusing on the delivery and collection of goods but involves also in different application areas arising from the transportation and logistics system [16].

Different VRP models have been developed for different applications to study the Routing Problem [1, 22]. The most recent is the one that has been devoted to more complex variants of the VRP occasionally called "rich" VRPs. These are closer to the VRP models [23] and are used to design an optimum allocation system that can improve the level of their services [27]. Toth Paolo and Vigo Daniele [16] have reported that the use of computerized methods in the VRP problems has significantly saved the computational effort ranging from 5 to 20%.

VRP can be designed as a directed or undirected graph subject to the problem environment [21, 28]. In the case of classical VRP where customers or customer's demands are known in advance and the driving time between the customer and the service time at each customer are also known, it can be defined and formulated in this section as a general VRP model [28].

It is designated in graph theory. To define the general model of VRP, let $G = (v, A)$ be an asymmetric graph where $V = \{0, 1, \dots, n)$ is a set of vertices representing cities or depot situated at the vertex 0, and A is the set of arcs. In every $\text{arc}(i,j); i = j$ is related to a nonnegative distance matrix $C = (C_{ij})$. In some cases, C_{ij} can be understood as a travel cost or as a travel time [21, 29, 30], it is often appropriate to substitute A by E. Where E is a set of undirected edges in the graph to represent asymmetric or undirected graph.

The common VRP comprises a set of at most K delivery or collection routes plan such that each route starts and ends at the depot, each customer is visited exactly once by exactly one vehicle, the total demand of each route does not exceed the vehicle capacity and the total routing cost is minimized. With all these assumptions, according to Stewart and Golden [31], a compact representation of VRP can be presented as follow:

$$\text{Minimize} = \sum_{k=1}^{n} c_{ij} x_{ij}^{k} \tag{1}$$

Subject to

$$\sum_{i,j}^{n} q_i x_{ij}^{k} \le Q, k = 1, 2, ..n \tag{2}$$

Where:

c_{ij} = The cost/distance of traveling, from i to j.

x_{ij}^k = 1 if vehicle k travels from i to j; 0 otherwise

m = The number of vehicles available

S_m = The set of all feasible solutions in the m-traveling salesman problem (m-TSP)

q_i = The amount demanded at location i

Q = The vehicle capacity.

In many realistic cases, the cost or the distance matrix satisfies the triangular inequality such that Eq. (3):

$$c_{ik} + c_{kj} \ge c_{ij}; \forall_{i,j,k} \in V. \tag{3}$$

In the VRP models, a differentiation has to be made between symmetric Eq. (4) and asymmetric Eq. (5). Solution approaches can vary significantly between these two cases [30].

$$A = \{(i,j) \mid i \in V, j \in V, i = j\} \tag{4}$$

$$A = \{(i,j) \mid i \in V, j \in V, i < j\} \tag{5}$$

In the real world, however, the general VRP model is enhanced by various constraints or side-constraints, [5]. The constraints can be such as vehicle capacity or time interval in which each customer has to be served [16], revealing the Capacitated Vehicle Routing Problem (CVRP) [20] and the Vehicle Routing Problem with Time Windows (VRPTW) [18, 21].

VRP models, whether they are used for public transport or transit, as well as distribution and logistics, they share certain mutual features. That is, they focus on the optimization of cost (working cost), distance covered, waiting time, etc.

Route No.	Origin(local Name)	Destination (local Name)	No. of busses	Kms
1	Megenagna	Kara	2.00	7.70
2	Kore Mekanisa	Addis Ketema	4.00	11.10
3	Ayertena	Minilik Square	8.00	10.80
4	Haliti	Addis Ketema	5.00	19.40
93	Bole Bulbula	Megenagna	2.00	15.20
	Total		**36.00**	**132.80**

Table 1.
The fixed number of busses assigned on selected routes (as of 2011).

4. Methodology

The current operations schedule of ACBSE has faced many challenges in its schedule busses assignment for each route. The enterprise uses a scheduling system that operates from 6:30 to 20:30. As shown in **Table 1** below, The enterprise has a fixed scheduling system, which means, the number of busses allowed on each route remains constant irrespective of the commuters' demand. In reality, however, the number of busses in each route should have been varied and can be determined based on the commuters' demand distribution. These approaches have made some busses moving empty while others being congested and resulting in capacity loss to the enterprise and bad service to passengers.

The enterprise occasionally involves additional busses on few routes during peak hours to overcome the challenges. Due to the absence of records for such decisions, therefore, in this study, the fixed number of busses assigned on each route is only considered as a basis for benchmarking.

5. Data collection

The data were collected only on the 93 routes that have been used by the enterprise for more than a decade. The data includes route performances, number of passengers served per route, the total trips made, revenue generated, running cost, and distance traveled. Moreover, data were also collected concerning the number of busses, bus capacity, average bus travel time, the length of each route, and working hours. The data were analyzed and organized per shift per route basis for validation purposes. The operating shifts, the time interval for each shift, and the passengers demand proportion per shift are reported in **Table 2**.

As per the report of the company, the enterprise has 534 busses and transported about 640, 000 passengers per day in 125 routes. In addition to these, the enterprise covers about 138 km and makes 61.50 trips per day [32]. In this study, it is also possible to classify the enterprise schedules in four shifts, i.e. peak hour and off-peak-hour schedules. These are two peaks and two off-peak periods during the services time (from 6:30 to 21:00). The morning peaks are from 6:30–9:30 and 15:30–19:30, while the off-peaks are also in the morning and afternoon shift. The details are reported in **Table 2**.

The demand proportion has also distributed within the four shifts. In the morning peak hour, the enterprises serve about 40% of the daily passengers. In the second peak-passengers' demand occurs in the evening shift shares about 35% of total passengers. The other periods are the first and the second off-peak shift which

Shift	Time interval	Duration (minute)	Demand proportion (%)
Morning peak hours	6:15–9:30	195	40%
First off-peak hours	9:30–15:30	360	20%
Evening peak hours	15:30–19:30	240	35%
Second off-peak hours	19:30–21:00	90	5%
Total		870	100

Table 2.
Demand proportion and duration of each shift.

occurs after the morning peak and the evening peak shift in which 20% and 5% for the demand proportion is allocated respectively.

6. Model development

The researchers have investigated the existing operating systems of the ACBSE, and a linear programming (LP) model was developed. The model helps to achieve a solution for the scheduling problems of the enterprise. The LP model developed in this study is a new approach in the literature of VRP that considers the trips made by two different types of busses to address the demand distribution of passengers in 93 routes in four working shifts.

6.1 Running the model

The LP model was fitted with data. To achieve its object, the model was coded and run using the General Algebraic Modeling System (GAMS) optimization software. The GAMS code was running within 0.15 seconds on 3.10 GHz, Window7 Home Premium, 4GB RAM, and core (i5) Dell Personal computer (Optiplex 790-Model). The resulting solution of the LP-model was first the number of trips per route per shift for each type of bus. After obtaining the solution, then, the number of trips was translated into the number of busses per route per shift for each type of bus.

6.2 Model validation

The LP model was investigated and validated to identify potential areas of improvement in the scheduling and assignment problem of ACBSE. It is also used to determine the number of busses to be assigned in a given route at a given shift that can address the demand distribution of passengers each shift for the 93 routes. The results of the LP-model were validated by comparing the current schedule and performances of the enterprise. The validation was made using four performance measuring parameters namely bus utilization, the total number of trips made, the total distance traveled, and the different operating costs.

7. Develop the LP model

In order to develop the LP mode, leti be the number of trips per route per shiftj made by bus type-I is represented by "x_{ij}" and by that of bus type-II is marked by "y_{ij}". The model is developed as a general model, at this section, then later is fitted to the ACBSE problem with data collected from the enterprise.

Definition of terms:
i = Route number, where $i = 1, 2, 3, \ldots, n$
j = Shifts, $j = 1, 2, \ldots, m$
D_{ij} = Passenger demand of route i at shift j
C_x and C_y = Capacity of bus type-I and bust type-II, respectively
M_j = Minimum Number of trips required at a given shift j
F_x and F_Y = Total available busses type-I and bus type-II, respectively
T_{ij}= Trip factor, maximum trips a bus can be made on route i per shift j
P_i = Trip proportion for route i

The objective is to minimize the total number of trips made by bus type-I and bust type-II. This objective is represented by Minimize : $\sum_{j=1}^{m}\sum_{i=1}^{n}\left[x_{ij} + y_{ij}\right]$ but needs to fulfill other various constraints.

The first constraint is to set the overall bus capacity of the Enterprise. This is the capacity of the two types of busses assigned in route i during shift j. This capacity should be greater than or equals to the total demand required by passengers that require ta rip in route i at shift j. It is Mathematically expressed as $C_x x_{ij} + C_y y_{ij} \geq D_{ij}$. The second and third constraints are used to check the total number of busses for each type of bus. To this effect, the total trip required by bus type-I and bus type-II should be less than or equal to the total trips available by bus type-I and bus type-II respectively. These are Mathematically expressed as: $\sum_{j=1}^{m}\sum_{i=1}^{n} x_{ij} \leq F_x \sum_{j=1}^{m}\sum_{i=1}^{n} T_{ij}$; and $\sum_{j=1}^{m}\sum_{i=1}^{n} y_{ij} \leq F_Y \sum_{j=1}^{m}\sum_{i=1}^{n} T_{ij}$.

The fourth constraint will determine the total minimum number of trips required in every 30 minutes for each of the 93 routes. This constraint is given b, $\sum_{j=1}^{m}\sum_{i=1}^{n}\left[x_{ij} + y_{ij}\right] \geq 93\sum_{j=1}^{4} M_j$. The trip made by bus type-I and bus type-II for each route i at each shift j should also be less than the total proportion of the trip available. This constrain is mathematically stated as: $x_{ij} \leq F_x P_i T_{ij}$; and $y_{ij} \leq F_y P_i T_{ij}$. The last constraint that shows sum of proportion. It must be summed and become one. This is written as $\sum_{i=1}^{m} P_i = 1$.

After compiling all the above constraints, the overall LP-model that determines the optimum number of trips i required per route per shift j is formulated as follows.

$$\text{Minimize} : \sum_{j=1}^{m}\sum_{i=1}^{n}\left[x_{ij} + y_{ij}\right] \tag{6}$$

Subject to

$$C_x x_{ij} + C_y y_{ij} \geq D_{ij} \tag{7}$$

$$\sum_{j=1}^{m}\sum_{i=1}^{n} x_{ij} \leq F_x \sum_{j=1}^{m}\sum_{i=1}^{n} T_{ij} \tag{8}$$

$$\sum_{j=1}^{m}\sum_{i=1}^{n} y_{ij} \leq F_y \sum_{j=1}^{m}\sum_{i=1}^{n} T_{ij} \tag{9}$$

$$\sum_{j=1}^{m}\sum_{i=1}^{n}\left[x_{ij} + y_{ij}\right] \geq \mathrm{m}\sum_{j=1}^{m} M_j \tag{10}$$

$$x_{ij} \leq F_x P_i T_{ij} \tag{11}$$

$$y_{ij} \leq F_y P_i T_{ij} \tag{12}$$

$$\sum_{i=1}^{m} P_i = 1 \tag{13}$$

$$x_{ij}, y_{ij} \geq 0 \tag{14}$$

The objective function 6 will minimize the total number of busses required to serve the total demand. Constraint 7 represents the combined capacity of busses assigned to each route i at a given shift j. Eq. 8 and 9 represent the total number of busses that need to be assigned for each route. The total required number of busses must be equal or less than the available number of busses the enterprise has; Eq. 10 shows that the total number of busses to be allotted on each route per shift has to fulfill the minimum number of busses required per route per shift. Eq. 11 and 12 are the number of busses to be assigned to a given route. Moreover, Eq. 13 warrants that the total sum of a probability is always one; last Eq. 14 is nonnegativity constraints.

This model guarantees at least one round trip for each route every 30 minutes to maintain a quality service level. It also contributes to the scientific body of knowledge by introducing different bus types in a single LP-model as a new constraint.

In the model, the travel time of bus on a given route was considered as the total sum of passenger boarding and alighting time (dwell time), acceleration and deceleration at bus stops, traffic light, and transfer time between each stop. However, little attention was also given to consider the functionality of the model and its output. In particular, the model was run once for all of the four shifts. The number of busses will be checked to make sure, M_j is met for all time slots. The results obtained may be fractional but rounded up later into upper integer values. All these may affect real mathematical optimality.

7.1 Minimum number of buses

For clarity and understanding, moreover, some of the parameters were defined. Some of the parameters are, M_j, T_{ij}, and P_i. They are explained and illustrated in detail below. The value of M_j is the minimum number of trips required at shift j to meet the maximum allowable waiting time. In this model, a single trip time for a given shift is 30 minutes in the length of the period which is 4 hours. For a given shift, then a minimum of 8 trips are required to limit the maximum waiting time of passengers to 30 minutes. This means, there should be at least one bus every 30 minutes or half an hour for each shift to provide quality service. The actual minimum number of busses required for this model is one bus because within a trip time of 30 minutes one bus can make eight trips during a 4 hour time interval. Therefore, M_j is given by:

$$M_j = \frac{Total\ Duration\ for\ shift\ j\ (minutes)}{30\ minutes} \tag{15}$$

Thus using Eq. 15 and the data reported in **Table 2** above, the value of M_j for the four shifts are computed and presented in **Table 3**.

7.2 Trip factor analysis

The other parameter that needs to be defined and explained is the trip factor, T_{ij}. It is computed using Eq. 16. This value is the minimum number of trips a bus can make on route i at a given shift j. Since the model computes the total trips that are

Route No.	Demand	Demani (D_{ij}) per Shift					Trip factor (T_{ij}) per Shift			
		Pi	D_1 (m_1 = 7)	D_2 (m_2 = 12)	D_3 (m_3 = 8)	D_4 (m_4 = 3)	1	2	3	4
1	4126	0.014	1650	825	1444	206	7	12	8	3
2	3497	0.012	1399	699	1224	175	4	7	5	2
3	11,030	0.038	4412	2206	3860	551	4	7	5	2
4	3029	0.010	1212	606	1060	151	3	5	3	1
93	778	0.003	311	156	272	39	2	4	3	1

Table 3.
Input parameters for the LP model for some routes.

needed per route per shift, the trip factor is used to calculate the available number of trips per route per shift.

The trip factor is the maximum number of trips a bus can make on route *i* per shift *j*; this factor is used to get the available number of busses in terms of trips. This is because the model computes the total number of trips that are required per route per shift. The actual number of busses is, then, calculated from the trip factor by setting how many trips a single bus can make at a given period.

$$T_{ij} = \frac{Total\ Duration\ for\ shift\ j}{Single\ trip\ travel\ time\ for\ route\ i} \tag{16}$$

By multiplying the trip factor by the available number of busses, that is F_x and F_y, it can help to find the maximum possible trips made by the total available busses.

7.3 Trip proportion

The other parameter that needs explanation is the trip proportion, P_i. It is required for route *i* during shift *j*. P_i is the trip proportion of a given route *i* from the overall routes also given by Eq. 17? P_i is used to determine trips to route *i* from the total available. The number of trips for each route is also computed based on the proportion of the total trips of all the routes. It is given by the following equation.

$$P_i = \frac{Daily\ Demand\ of\ Route\ i}{Total\ Daily\ Demand\ of\ all\ Routes} \tag{17}$$

The last parameter the requires explanation is D_{ij}. It is the average daily passengers of route *i* during shift *j* that requires transport services. It is collected from the secondary data sources of the enterprise. It is allocated per shifty by multiplying the average number of daily passengers of route *i* by the demand proportion of shift *j* and is reported in **Table 2**.

7.4 Solve the model

The next step is to run the model to obtain feasible solutions. The LP model is solved based on the data of the average daily passengers that have been transported for the last 19 months in four shifts. The daily passengers' that demand for transport for the last 19 months was collected and then the average daily passenger' demand of each month was computed per route per shift based on the trip proportion (P_i) of route i, and reported in **Table 3**.

7.5 Input parameter to the model

In the process of running and solving the model, first, the input data has to be fitted. In this regard, to fit the LP-model with the input parameters that are involved in the model, first it the parameters needs to be determined. These parameters are either computed or collected from the enterprise. The sample input parametric values are shown in **Table 3**.

These inputs parameters are standard carrying capacity of busses, the operational number of busses, the passenger that demand transport services per route per shift (D_{ij}), the trip factors (T_{ij}), the minimum number of trips per shift (M_j) and the trip proportion per route (P_i). The sample input parametric values of a few routes are shown in **Table 3**.

There are four types of busses used by the enterprise (namely DAF, Mercedes, Single, and rigged Articulated busses) but they can be categorized in two based on their seat capacity. These are one bus with seat capacities of 30 passengers (DAF, Mercedes, Single, and rigged Articulated) and busses with seat capacity 50 passengers (the Articulated one).

While fitting to the LP-model busses with a seating capacity of 30 are classified as bus type-I (but can transport 60 passengers) and busses with a seating capacity of 50 are classified as bus type-II (but can transport 90 passengers). The maximum number of capacity, 60 passengers and 90 passengers are based on the standard capacity of public bus transportation [33]. The total capacity of each bus type is equal to the seating capacity plus the standing capacity. The enterprise has a total number of type-I and type-II is 600 and 90, respectively. Thus, the objective function of the research is used to compute the optimum trips and mixes of the two types of busses per route per shift.

The total operational busses in bus type-I are 600 and that of bus type-II is 90 busses. The numbers of operational busses are not only 690, but the rest of the operational busses are kept for backups during failure and other services such as contract and employee service. Also, the 93 routes which are under analysis serve more than 90% of the demand during a day and thus the operational bus assignment is based on this proportion. After substituting the values of input parameters and constants into the LP model, the model can be re-written as:

$$\text{Minimize}: \sum_{j=1}^{4} \sum_{i=1}^{93} \left[x_{ij} + y_{ij}\right] \tag{18}$$

Subject to

$$60x_{ij} + 90y_{ij} \geq D_{ij} \tag{19}$$

$$\sum_{j=1}^{4} \sum_{i=1}^{93} x_{ij} \leq 600 * \sum_{j=1}^{4} \sum_{i=1}^{93} T_{ij} \tag{20}$$

$$\sum_{j=1}^{4} \sum_{i=1}^{93} y_{ij} \leq 90 * \sum_{j=1}^{4} \sum_{i=1}^{93} T_{ij} \tag{21}$$

$$\sum_{j=1}^{4}\sum_{i=1}^{93}\left[x_{ij}+y_{ij}\right] \geq 93 * \sum_{j=1}^{4} M_j \tag{22}$$

$$x_{ij} \leq 600 * P_i T_{ij} \tag{23}$$

$$y_{ij} \leq 90 * P_i T_{ij} \tag{24}$$

$$\sum_{i=1}^{93} P_i = 1 \tag{25}$$

$$x_{ij}, y_{ij} \geq 0 \tag{26}$$

8. Model output

The output of the model indicates the total number of trips by the two types of busses needed to serve the average demand of each route on a given shift. The sample outputs of the LP model are shown in **Table 4**.

The GAMS build system has different solvers such as BARON, BDMLP, and BENCH. Replace this text with your section Heading CNOPT, CPLEX, LGO, etc. that are capable of solving different varieties of problems. After trying each solver, for reporting purposes, CPLEX solver is chosen to solve the LP model developed above. The LP model is coded and programmed using the GAMS. This build system and the piece of the GAMS code are reported in the Appendix.

The outputs of the model are reported by taking the upper integer value. As shown in **Table 4**, for example for route 3, 61 trips by bus type-I and 14 trips by bus type-II were required for shift one. There are also routes where no trips are required by bus type-I in the off-peak shifts. Since the LP model produces the number of trips required, the output has to be converted into the number of busses required for each route in a given shift. This has to be done by dividing the number of trips

Route No.	Shift 1		Shift 2		Shift 3		Shift 4	
	Bus type-I	Bus type-II	Bus type-I	Bus type-II	Bus type-I	Bus type-II	Bus type-I	Bus type-II
L	19	9	0	10	14	11	0	3
2	19	5	5	8	15	6	0	2
3	61	14	14	24	48	17	0	7
4	18	3	6	5	15	3	2	1
5	14	3	4	6	12	3	0	2
6	51	12	0	21	41	14	0	6
.	.	.	.	.	.	.	.	.
.	.	.	.	.	.	.	.	.
.	.	.	.	.	.	.	.	.
91	4	1	2	2	4	1	1	1
92	21	5	0	9	17	6	0	3
93	11	4	0	5	9	4	0	2
Total	1541	402	174	618	1231	490	18	156

Table 4.
Number of trips required per route per shift.

from the model output by the trip factor (T_{ij}). The sample of the number of busses per route per shift presented in **Table 4** and computed for all 93 routes in the same way.

9. Assigning buses to routes

Based on the results of the output of the LP-model, there are 4627 total trips required to serve the average daily passenger demand. The actual number of busses required for a given route i during a given shift j depends on the trip factor. That is the number of trips that can be made by each bus type during a given shift. Thus, the output needs to be transformed into the number of busses needed for each route per shift based on the feasible trips that a bus can make during a given shift j on each route i. From the output of the LP model shown in **Table 3**, the number of trips can be transformed into number busses using the following equation.

$$Number\ of\ Bus = \frac{Number\ of\ Trips}{T_{ij}} \tag{27}$$

For some routes, the output of the LP is small for some route, so adjusting the actual number of busses is required for such routes to have at least two busses on a given route per shift to allocate them on both ends of the route; that is one on the going trip and the other on the returning trip. This is because the demand during a given shift j on each route i lies in both directions of the route. Sample of the number of busses per route per shift reported in **Table 5** and computed for all 93 routes in the same way.

The results showed that the actual number of busses required during peak periods is higher than that of off-peak periods. Thus, some of the busses that operate during the morning peak period have to wait on bus stops until they are required for the evening peak.

	Shift 1		Shift 2		Shift 3		Shift 4	
Route No.	**Bus type-I**	**Bus type-II**	**Bus type-I**	**Bus type-II**	**Bus type-I**	**Bus type-II**	**Bus type-I**	**Bus type-II**
1	3	1	1	1	2	1	1	1
2	5	1	1	1	3	1	1	1
3	16	3	2	3	10	4	0	3
4	6	1	2	1	4	1	2	0
5	4	1	1	1	3	1	1	1
6	12	3	0	3	8	3		
.	.	.	.	.	.	.	.	.
.	.	.	.	.	.	.	.	.
.	.	.	.	.	.	.	.	.
91	3	0	2	0	3	0	3	0
92	5	1	1	1	3	1	1	1
93	2	1	1	1	1	1	2	0
Total	396	94	114	88	269	94.8	121	79

Table 5.
Number of busses per route per shift.

Similarly, the actual number of busses required for each shift varies and the number of busses required during peak periods is higher than that of off-peak periods. Thus, some of the busses that operate during the morning peak period have to wait on bus stops until they are required for the evening peak.

10. Model validations

The outputs of the LP-model were evaluated using different performance measuring parameters. For the validation purpose, various assessments were made between the existing performances and the LP improved systems. The comparisons made, in this research, were based on bus utilization, distance and trips covered and the different operating costs of the enterprise [32]. Each of them is discussed in the following sections.

10.1 Bus utilization

After converting and assigning busses to each route and each shift, then the improvements achieved by the LP model were compared with the existing bus utilization of ACBSE [32]. The bus utilization was calculated as the ratio of the number of passengers served by the bus to the number of passengers carrying capacity. The average daily bus utilization of the existing and the improved systems are presented in **Figure 3**. The results of the study show that the improved system has better bus utilization than the existing one. The existing system has a maximum bus utilization of about 125% on daily basis, which is very congested and crowded, while the improved system by this study has shown that a maximum bus utilization of 97% [32]. This shows how passenger congestion and service quality were improved by the new system. The average bus utilization for the improved systems is about 66.33% which is better than the existing systems that are 64.26%. Though the average utilization of both of the systems seems close to each other, in the improved most of the utilization lies between 60% and 80% whereas in the existing system had unbalanced bus utilization which is sometimes failed below 20% and other times above 120%.

As presented in **Figure 4**, an in-depth analysis of bus utilization based on shift was carried out. The improved system exhibited significant improvement as

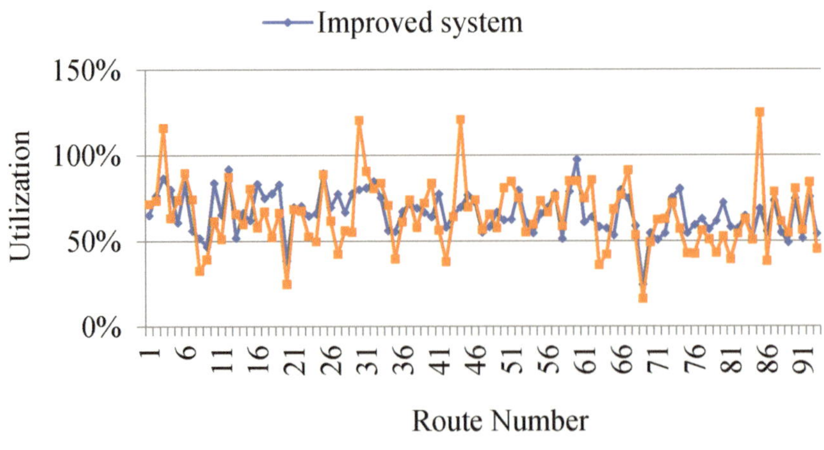

Figure 3.
Average daily bus utilization; current vs improved system.

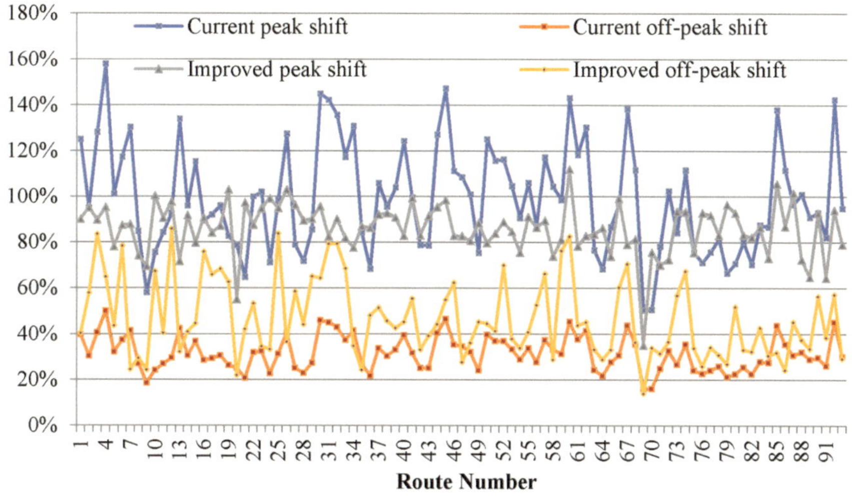

Figure 4.
Average bus utilization for peak and off-peak shifts; current vs improved system.

compared to the existing one and reported by shift. The improved bus utilization by shift is 89.8% during the morning peak, 51.19% during the first off-peak, 82.24% during the evening peak, and 42.1% during the second off-peak periods. The improvement as compared to the existing systems is 19.75% during first off-peak and 12.15% during the second off-peak. The bus utilization of the improved system also reduced the passengers' congestions in peak hours, i.e. bus utilization is decreased from 116.1% to 89.8%. Thus, the improved system has a relatively stable and consistent bus utilization with improved service quality. The new model also indirectly improves the service quality for passengers by reducing over crowdedness while during peak periods and reduces the operating cost during off-peak periods to the enterprise.

10.2 Distance and trip coverage

In this section, the distance covered by the existing and the new system are compared. To compare this, the total kilometer covered by busses for route i per day is computed. It is the sum of the Kilometer covered during all the four shifts. **Figure 5** shows that the total distance coverage of the current and improved systems. The distance covered on each shift was computed by multiplying the number of busses allocated to a given route at that shift by the number of trips that can be made by a single bus and by the route length. The total distance covered per day for the improved system is 70,964 kilometers, while for the existing system, it is about 78,963.7 kilometers per day. This shows a reduction of 10.13% in the daily distance coverage to serve the same number of passengers.

In this case, the total daily trips made for the improved system is also computed for each shift by multiplying the number of busses assigned and the number of trips a single bus can make during that shift. The total trips covered for the existing systems were 5504 trips per day while 5584 for improved systems. This also shows an improvement of 80 trips per day compared with the existing system. The increased number of trips was achieved with a 10.13% reduction on the daily Kilometer. This also improves service availability in addition to saving on the operating costs.

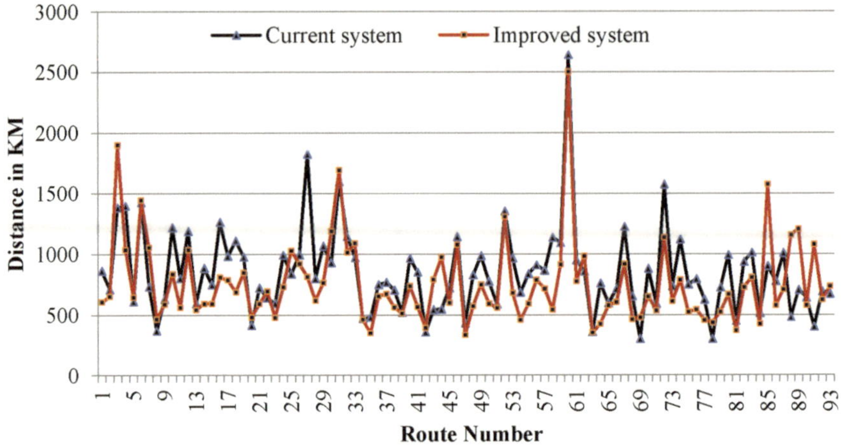

Figure 5.
Distance coverage; current vs improved system.

10.3 Operating costs

Using the different operating costs of the enterprise, the other performance measuring parameters were also assessed in this research. The improvements made by the new model were also considerable. The total daily operating cost of the enterprise for each route is the sum of operating costs for all the shifts. From the comparison made, the results of the study show that the average daily operating costs of the enterprise for the existing systems are about 1,101,283.68 ETB (ETB = Ethiopian Birr and 1USD = 34.33ETB as of Jun 4, 2020) while for the improved systems is 949,991.49 ETB. The saving of the new system in this read is about 13.74% per day compared to the current system. **Figure 6** shows the improvements made by the new systems are achieved nearly in all the operating costs of the enterprise compared to the existing one.

As compared to all the operating costs, larger saving is observed on gas oil. This reduction has also a strong relationship with the total Kilometer covered by each

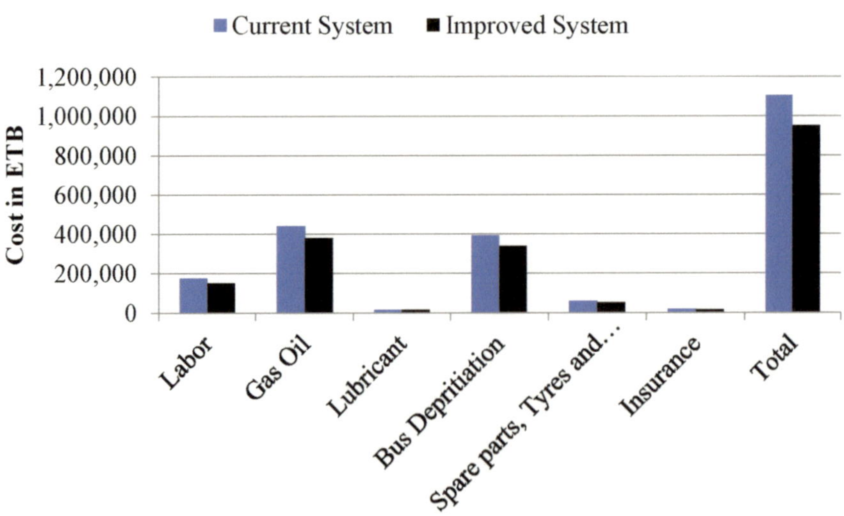

Figure 6.
Current vs improved operating cost.

bus. This is due to the fact that the total kilometer covered is improvement resulted in a reduction in the cost of gas oil consumed.

11. Conclusion

The motivation of this research was to develop a model that can optimize the operational performances of ACBSE. Based on the major findings, it can be concluded that the existing scheduling systems of ACBSE have shown low performances on the bus utilization, operating cost, and daily trips and distance covered. These improvements have been achieved because the existing system has fixed numbers of busses assigned to routes without considering the variability of passenger demands. This had cost more the enterprise. However, the operational performance improvements of the LP-model have shown better performances over the existing one nearly in all of the above performance measuring parameters. Besides, it can be concluded that the existing bus scheduling and operations system has a lower average utilization of busses compared with the new system by 2.1%. The bus utilization per route per shift also shows significant improvement over the existing system. With regard to the cost-saving, the new model has resulted in a saving of 13.74% (151,292.19 ETB) in the operating costs of the enterprise. Moreover, the new model also resulted in a 10.13% saving on the total km covered with 80 additional available trips per day for the enterprise. In addition to these and the saving in all the parameters, the improved system has also reduced the waiting time, improve service quality, and reduce passenger congestion by scheduling busses based on the international standard bus capacity. The new system has also a significant reduction in the total kilometer covered while improving the total trips made daily. All these improvements of the new system of the LP-Model were exhibited without altering the existing routes used by the enterprise. But rerouting the existing route's design may also bring radical improvement to the performances of the enterprise.

Appendix: the GAMS sample code

****

Defining Route i and Shift j

Sets i Routes /1 * 93/ This sets routes from route 1 to route 93.

j Time periods /1 * 4/; Sets shifts from shift 1 to shift 4.

Sample Demand distribution per route per shift

Table d(i, j) demand distribution on route *i* on time shift *j*

	1	2	3	4
1	2249	1124	1968	281
2	2249	1124	1968	281
3	2249	1124	1968	281
.	.	.	.	.
.	.	.	.	.
.	.	.	.	.
92	1848	924	1617	231
93	1362	681	1192	170

Sample Trip Factors (T_{ij})

Table T(*ij*) trip factor on route *i* for time period *j*

Minimum Trip Required (M *j*)

Parameters M(*j*) minimum trips required at time period *j*

Trip Proportion (P_i)

	1	2	3	4
1	7	12	8	3
2	4	7	5	2
3	4	7	5	2
.	.	.	.	.
.	.	.	.	.
.	.	.	.	.
92	8	5	2	
93	5	9	6	2

/1 7
2 12
3 8
4 3/

P(ᵢ) demand proportion of route i from all routes /

1 0.014
2 0.012
3 0.040
. .
. .
. .
92 0.012
93 0.009/;

Decision Variables

X(ij) Number of Bus type-I for route i during shift j
Y (ij) Number of Bus type-II for route i during shift j;
Variable Z; For the objective function

Model Equation

obj.. Z=e=sum((i,j), X(i,j))+ sum((i,j), Y(i,j));
*const1.. sum((i,j), X(i,j))*60 + sum((i,j), Y(i,j))*90 =g= sum((i,j),d(i,j));*
*const2.. sum((i,j), X(i,j))=l= sum((i,j), T(i,j))*600;*
*const3.. sum((i,j), Y(i,j))=l= sum((i,j), T(i,j))*90;*
const4.. sum((i,j), X(i,j))+ sum((i,j), Y(i,j)) =g= sum(j, M(j));
*const5.. sum((i,j), X(i,j))+ sum((i,j), Y(i,j))=l=sum((i,j),T(i,j))*690;*

Since bus type-II has greater capacity, priority is given so that demand is to be served by the available capacity of Y for that route to reduce number of buses required and if it is beyond the available number of Y for that route then the rest is to be served by the available number of X. Thus, based on this assumption the following equation sets the lower bound to values of X and Y, if not since it is minimization it starts from zero and can't display the appropriate mix of X and Y.

$Y.lo(i,j)\$[(d(i,j)/90) \leq (T(i,j) * P(i) * 90)] = d(i,j)/90;$ *lower bound for Y*
$Y.lo(i,j)\$[(d(i,j)/90) \geq (P(i) * T(i,j) * 90)] = P(i) \ *T(i,j) * 90;$
$X.lo(i,j)\$[(d(i,j)/90) \geq (P(i) * T(i,j) * 90)] = (d(i,j)/60) - (P(i) * T(i,j) * 90);$
lower bound for X
$X.lo(i,j)\$[((d(i,j)/60)-(P(i)*T(i,j)*90)) \geq (P(i)*T(i,j)*600)] = (P(i)*T(i,j)* 600);$

Thus from this, the assignment of bus type X depends on the assignment of Y and X is

assigned for a given route if the expected demand during that time period is beyond

the capacity and availability of bus type Y

Model eq / All / ;

```
Solve eq using lp minimizing Z;
display X.l;
display Y.l;
*****************
```

Author details

Eshetie Berhan* and Daniel Kitaw
Addis Ababa University, Addis Ababa Institute of Technology, School of Mechanical and Industrial Engineering, Addis Ababa, Ethiopia

*Address all correspondence to: berhan.eshetie@gmail.com

References

[1] Cordeau J-F, Gendreau M, Laporte G, Potvin J-Y, Semet F. A guide to vehicle routing heuristics. Journal of the Operational Research Society. 2002;**53**: 512-522

[2] Reimann M. Analyzing a vehicle routing problem with stochastic demand using ant colony optimization. In: EURO Working Group on Transportation. 2005

[3] Bunte S, Kliewer N. An overview on vehicle scheduling models. In: Decision Support & Operations Research Lab. Warburger, Germany: University of Paderborn; 2009

[4] Seong P, Jae. Bus Network Scheduling with Genetic Algorithms and Simulation [Master of Science Thesis]. College Park: University of Maryland, Department of Civil and Environment Engineering, University of Maryland; 2005

[5] Caric T, Gold H. Vehicle Routing Problem. Croatia: I-Tech, I-Tech Education and Publishing KG; 2008

[6] Christopher J, Goodson. Solutions Methodologies for VRP with Stochastic Demand [Dissertation] Iowa; 2010

[7] Clara N, Rosemary B, Jeff L, Robert S. A Set Partitioning-Based Model for the Stochastic Vehicle Routing Problem. Technical Report. Bethlehem, PA: Texas State University and Lehigh University, 601 University Drive San Marcos, TX 78666 or 200 W. Packer Ave; 2006

[8] Reimann M, Doerner K, Hartl RF. D-ants: Savings based ants divide and conquer the vehicle routing problem. Journal of Computers & Operations Research. 2003;**31**(4):563-591. ISSN: 0305-0548

[9] Antonios T, Ioannis M. Stochastic single vehicle routing with a predefined customer sequence and multiple depot returns. European Journal of Operational Research. 2009;**197**:557-571

[10] Bianchi L. Notes on Dynamic Vehicle Routing, the State-of-the-Arts. Techniacal Report Idsia-05-01. Switzerland: Istituto Dalle Molle di Studi sull'Intelligenza Artificiale; 2000

[11] Anbuudayasankar SP, Ganesh K. Mixed-integer linear programming for vehicle routing problem with simultaneous delivery and pick-up with maximum route-length. The International Journal of Applied Management and Technology. 2008; **6**(1):31-52

[12] Dantzig GB, Ramser JH. The truck dispatching problem. Journal of management science. Management Science. 1959;**6**(1):80-91

[13] Laporte G, Louveaux FV, Hamme v L. An integer l-shaped algorithm for the capacitated vehicle routing problem with stochastic demands. Operations Research. 2002;**50**:415-423

[14] Tillman FA. The multiple terminal delivery problem with probabilistic demands. Transportation Science. 1969; **3**:192-204

[15] Baldacci R, Vigo D, Toth P. Exact Solution of the Capacitated Vehicle Routing Problem. USA: John Wiley & Sons, Inc.; 2010

[16] Toth P, Daniele V. The Vehicle Routing Problem, SIAM Monographs on Discrete Mathematics and Applications. Philadelphia, PA: Society for Industrial and Applied Mathematics, Society for Industrial and Applied Mathematics; 2002. pp. 19104-12688

[17] Winston W. Operations Research: Applications and algorithms (with CD-

ROM and Infotrac). 4th ed. Thomson Brooks/Cole; 2003

[18] Calvete HI, Gale C, Oliveros MJ, Sanchez-valverde B. Vehicle routing problems with soft time windows: An optimization based approach. Journal of Monograf del Seminario Matemco Garce Galdeano. 2004;**31**:295-304

[19] Drex M. Rich Vehicle Routing in Theory and Practice. Technical Report Lm. Mainz, Germany: Johannes Gutenberg University; 2011

[20] Moses C, Samir K, Balaji R. Algorithms for capacitated routing. Journal of SIAM Journal on computing. 2001;**31**(3):665-682

[21] Xiangyong L, Peng T, Leung S. Vehicle routing problems with-time windows and stochastic travel and service times: Models and algorithm. International Journal of Production Economics. 2010;**125**:137-145

[22] Chenghua S, Xiaofeng Z. Research on model fitting capacity of vehicle routing problem. International Journal of Advancements in Computing Technology(IJACT). 2011;**3**(11):185-193

[23] Roberto B, Maria B, Daniele V. Routing a Heterogeneous Fleet of Vehicles. Technical Report deis. or. ingce 2007/1, DEIS, University Bologna, via Venezia 52, 47023 Cesena, Italy; 2007

[24] Laporte G. Fifty years of vehicle routing. Transportation Science. 2009; **43**(4):408-416

[25] Avishai C. Optimal multi-vehicle type transit timetabling and vehicle scheduling. Procedia—Social and Behavioral Sciences. 2011;**20**(0):19-30

[26] Baldacci R, Battarra M, Vigo D. Routing a heterogeneous fleet of vehicles. In: Golden B, Raghavan S, Wasil E, editors. The Vehicle Routing Problem: Latest Advances and New Challenges, Volume 43 of Operations Research/Computer Science Interfaces. US: Springer; 2008. pp. 3-27

[27] Mariam F, Ahmed M, Anouck G. Vehicle routing problem instances: Application to multi-uav mission planning. In: Conference, American Institute of Aeronautics and Astronautics (AIAA) Guidance, Navigation. Toronto, Ontario Canada; 2010

[28] Linong C-Y, Wan I, Khairuddin O, Zirour M. Vehicle routing problem: Models and solutions. Journal of Quality Measurement and Anlysis. 2008;**4**(1): 205-218

[29] Cordeau J-F, Laporte G, Mercier A. A unifid tabu search heuristic for vehicle routing problem with time windows. Journal of Operations Research Society. 2001;**53**:928-936

[30] Wouter J. Approximate Models and Solution Approaches for the Vehicle Routing Problem with Multiple Use of Vehicles and Time Windows [Master thesis] Middle East Technical University; 2008

[31] Stewart WJ, Golden BL. Stochastic vehicle routing: A comprehensive approach. European Journal of Operational Research. 1983;**14**:371-385

[32] Berhan E et al. Performance Analysis on Public Bus Transport of the City of Addis Ababa. International Journal of Computer Information Systems and Industrial Management Applications. 2013;**5**:722-728. ISSN: 2150–7988

[33] TCRP. Transit Capacity and Quality of Service: Part 4 Bus Transit Capacity Manual. 2nd ed. Vol. 8. Orlando, FL: Kisttelson and Associates, Inc; 2012

Chapter 10

Impact of ICT to Improve of the Manufacturing in a SME Biomedical of Mexicali, Mexico

Roberto Carlos Valdés Hernández,
Juan Gabriel Lopez Hernandez,
Adelaida Figueroa Villanueva
and Vidblain Amaro Ortega

Abstract

This work presents a way to optimize the manufacturing processes in a small biomedical industry considered in the micro-, small-, and medium-sized enterprises (SME) group and located in the Mexicali city, using a specialized software that act as design and test of a new model, being the COSIMIR (Cell Oriented Simulation of Industrial Robots) software. With this software was designed a new industrial process in a workstation separated of the main step of a manufacturing line, where are fabricated biomechanical knees pads. The process was made as a manual activity in a work station and had to be separated from the conveyor belt of the main activities, because where previously made by an automatized device that was failing continually and was delaying the delivery to the next steps of the manufacturing processes and to the customers as a final product fabricated in this industry. In this place of the company, an operation was made to organize the biomechanical knee in a plastic container with divisions and to be transported safe and quickly to other area by a conveyor belt with linear process flow. The investigation was conducted from 2018 to 2019.

Keywords: ICT, SME biomedical industry, COSIMIR software, manufacturing processes

1. Introduction

The optimization in industrial process reduces and eliminates a lot quantity of failures and errors, generating more control in any step of the manufacturing areas. The majorly of the times are used specific improvements tools, as Ishikawa diagram and control graphs, principally; and in this times is used information and communication technologies (ICT) as complement of the optimization of the fabrication processes, with software to simulate the industrial activities, before applied in the industrial processes of any type of industrial company [1]. This has been to optimize the industrial operations and supports greatly to the industrial plants, where software have developed by specialized programmers, simulating the functionability of each step of the manufacturing processes and to have an idea of the industrial operations, as the possible fails and errors to avoid unnecesary costs [2]. One of the most important industries worldwide is biomedical, which manufactures various devices used in everyday life in the SME biomedical industry of the Mexicali city [3]. There are much software to simulate industrial operations with design and test functions, being one of most used in the industry for the design of equipment, machines and tools, the COSIMIR software. This software contains basic and specialized functions with illustrations are made for industrial certain operations. In this investigation, this software was used to design a new work table, being developed in a special step of a manufacturing line out of the conveyor belt where biomechanical knee pads were organized in a plastic container divided into sections. This evaluation was made because, in this industrial operation, they were presented high indices of deffective products manufactured, before be organized in the containers, for the rawhen processing them quickly as the operation of the conveyor belt. Due to this, the following operations on the conveyor belt could not be developed by workers, causing delays in delivery to the following areas and to the customer. This generated concern in managers and supervisory personnel, the solution being the implementation of a work table in an annex place at the conveyor belt where it previously was made. One aspect observed was the evaluation of development of operations with security and comfort, analyzing the movements and postures of workers in each operation of the new workstation as ergonomic way [4, 5]. This caused at first, a greater concern of the operating personnel by the possible presence of a fatigue and stress, by inadequate postures and quantity of movements but was solutionated by ergonomic methods, to elaborate the activities in a manual way adequately, with correct light as is showed in **Figure 1**.

1.1 Impact of ICT in the industry

Mexicali, which is located in the northwest of the Mexican Republic, is considered an industrial city where there are around 50 SME industrial companies of various types of manufacturing, which use the ICT [6, 7]. ICT are a very basic tool in industrial activities of big and SME industries, where they have infinity of software in various operations. In all industries, various analyzes are elaborated with software that support the improvement of functions, being mainly to develop

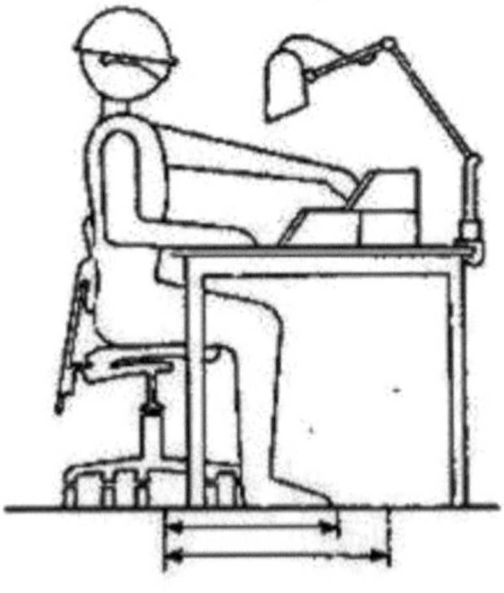

Figure 1.
Correct posture and installation of the new worktable proposed in the investigation (2018). Source: https://www.monografias.com/trabajos101/a-estaciones-trabajo-aburridas-solucion-aplicacion-ergonomia/a-estaciones-trabajo-aburridas-solucion-aplicacion-ergonomia2.shtml

simulation evaluations, which determine if it is possible to apply the pertinent improvements for the solution of problems [4]. This is evaluated to increase levels of certain parameters such as production and quality and to reduce or eliminate human errors or occupational risks [4]. When in an industry integrates to the use of the Internet, ICT are part of it, requiring specialized services and evaluating costs to obtain the greatest benefit from ICT and achieve optimum growth for the company [8]. The ICT have managed to impact the work centers in such a way that sometimes the interaction between workers in administrative and production areas is through mobile or desktop computers, and if there is no control over this, the communication between industrial equipment and machinery, is more difficult, originating sometimes fails and errors, as is showed in **Figure 2**. The ICT have successfully brought people together through videoconferencing, information exchange, and remote process assessment of industrial processes, which has greatly supported in the optimizing of operations and decision-making without being in the manufacturing areas. Another important aspect is to be able to easily and quickly obtain relevant information to the company, to elaborate the pertinent analysis and continuous improvement, to increase the operating performance of both workers and industrial equipment and machines from any place of the world [5].

1.2 COSIMIR software

This a very friendly and easy to learn software that was developed by the FESTO Company for analysis of robot operations, but as its users have specialized in this software, it has been used for other types of designs, such as the work table of this investigation [9]. A main feature of this software is to be able to add objects for any type of industrial or other activity to be used in industrial plants [10]. This company created this software to develop and evaluate simulations with robots in industrial processes and dedicated educational institutions to optimize the teaching learning processes, so that the students had notions at the time to begin to work. In this investigation, this software was applied to elaborate the development of new tools and structures to improve working conditions, increasing the production and

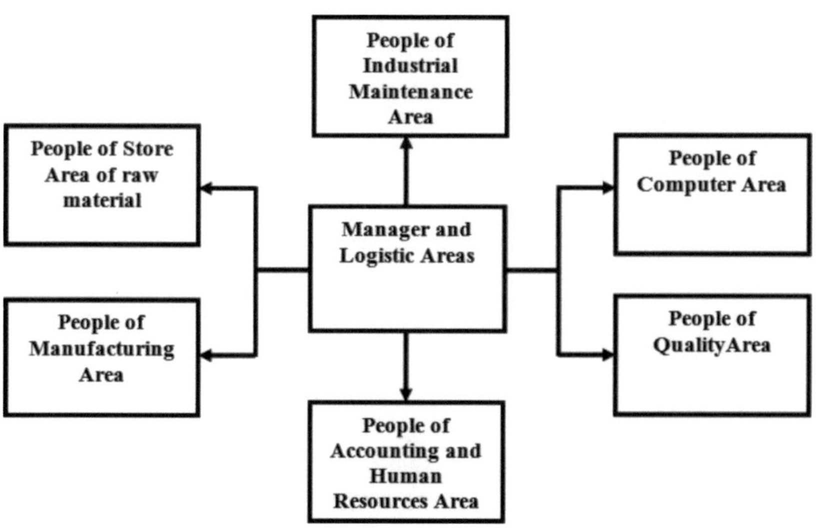

Figure 2.
Adequate communication using ICT in the SME biomedical industry evaluated.Source: Analysis of the investigation.

quality indices and reducing fails and errors. With this software, was made previous simulations very easy that supported to determine the types of adequate workstation to any type of workers. This modeling software allowed knowing the operation of new designs workstations and determine the most efficient, of by applying integrated manufacturing tools by computer that is showed in **Figure 3**. The ICT are very useful in the automation of industrial processes, such as the one used in this study, with a coupling system with the computer and a specialized database [11]. Manufacturing processes were found to improve productivity and quality rates with automated electronic systems controlled by ICT. The COSIMIR software that can be evaluated objects and figures in 3D, has been very utilized from the late twentieth century in educational activities and investigation laboratories, and in industries this software began to use in industries and in the last 10 years was increased its use. It is a friendly software to students and was generated a great interest to be utilized in some industrial activities. The ICT are very useful in the automation of industrial processes, such is showed in this investigation with a docking system between a computer system and a specialized database.

1.3 Computer integrated manufacturing

There are countless industrial processes that are the main structure of the activities of industrial plants, and are constituided of automated and manual operations, which use the Computer Integrated Manufacturing (CIM) processes. The functions that are performed manually are developed in assembly lines and in worktables, where they are grouped by manufacturing cells. In all industrial processes there are relevant factors that improve quality and tend to increase productivity, applying process optimization methods. This generates competitiveness in industrial companies. One factor is the safety and comfort of the operating personnel who elaborate

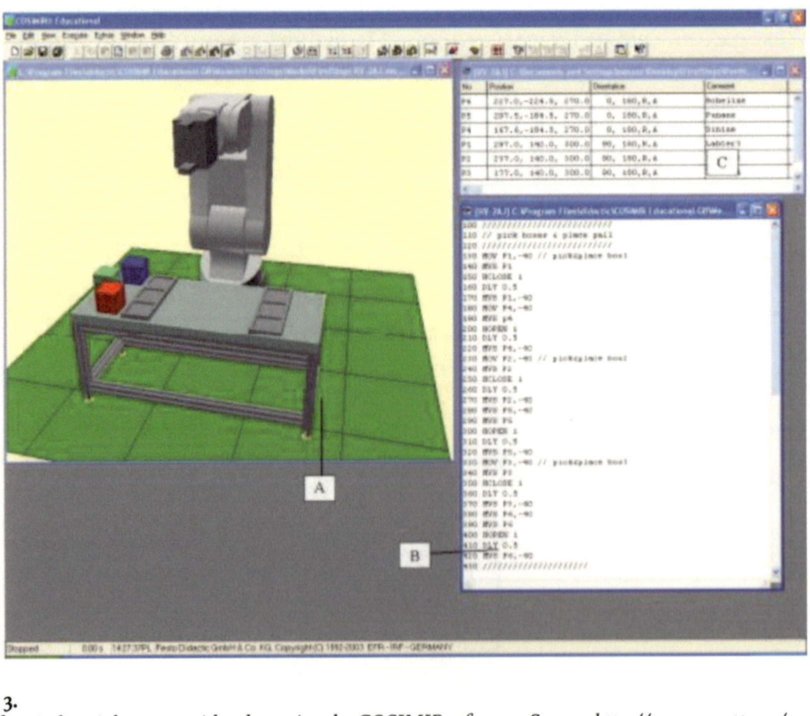

Figure 3.
View of an industrial process with robot using the COSIMIR software. Source: http://www.ene.ttu.ee/elektriajamid/oppeinfo/materjal/AAR0040/Pt31_COSIMIR_Educational.pdf

the activities, since musculoskeletal disorders are sometimes generated, caused by cumulative trama by inadequate postures and movements of workers. In this study elaborated, was suggested to evaluate the industrial processes in a work station and optimize them to obtain maximum operative yielding of industrial equipment and machinery, and of people that was working in manufacturing areas. For this reason, a new industrial process was collocated with a new worktable to improve the industrial process of the manufacturing line evaluated [11]. In this industry evaluated, there are areas with computers, from where are controlled industrial processes analyzing the information with the COSIMIR software to control functions of industrial machinery and equipment with automated functions. Also, manual operations elaborated by workers to evaluate their operational performance were evaluated, and improvements to be developed to optimize manufacturing processes were analyzed. Simulations with COSIMIR software are also generated in these areas. **Figure 4** illustrates an area of computer systems that control the industrial processes.

1.4 Optimization of industrial processes

The operation of industrial plants is made up of tasks with decision-making at the different hierarchy levels, such as planning and scheduling, in addition to optimization and control. Increasing competition from any type of industry requires an operation to have a more agile distribution of manufacturing areas, in order to increase productivity with flexible operations, generating a decrease in the total cost of production [12]. This requires optimization in various industrial processes of the manufacturing areas, applying two essential techniques such as real-time optimization (OTR) and optimization represented by a mathematical model in steady-state conditions and with linear or non-linear equations of industrial processes [13, 14]. In both cases, there are certain limitations to achievable flexibility and economic benefit, especially when considering the use of dynamic processes as continuous processes and with batch operations. In the industries it is common to use control systems based on the feedback of the output variables with the Proportional, Integral and Derivative (PID) controls. A representation of manufacturing processes is in **Figure 5**.

Figure 4.
Area which is controlled by computer the industrial processes. Source: Photography of manufacturing area of a biomedical industry in Mexicali.

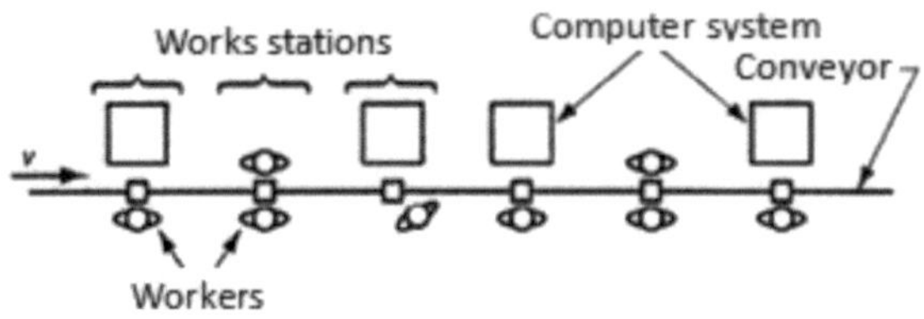

Figure 5.
Schematic diagram of manufacturing area in the electronic industry in Mexicali. Source: Analysis of the investigation.

1.5 Numerical analysis

Only in high level efficiency equipment, it is necessary to use the OTR, with analysis of processes in current periods and with future predictions. In this investigation was used the MATLAB software [15] with mathematical and statistical analysis, to develop the evaluations which the current conditions and the improvements necessary of the manufacturing processes for the optimal operational performance of the evaluated new workstation. Industrial operations are constantly evaluated by MATLAB software that indicates by means of numerical values or representations with signals of various figures in tables or graphs. The specialized personnel of this type of activities have the function of developing analyzing the information obtained from the operational performance of equipment, devices and industrial machinery and developing new prototypes based on the needs of the industries.

Ishikawa diagram used to evaluate industrial processes.

The Ishikawa diagram was is an important tool of continuous improvement and was utilized in this investigation, to evaluate the six parameters that are presented more frequently in any type of operations in any type of industry [16, 17] and in this case was analyzed each step of the process explained in the last section. The Ishikawa diagram is showed in **Figure 6**, and was used with six factors involved in each operation of the company evaluated, being the analysis of:

1. **People:** Evaluation of workers in the manufacturing line investigated, analyzing the time and movements in each operation.

2. **Environment:** Analysis of the relationship of management and supervision personnel with workers who elaborated the industrial activities in each step of the manufacturing line evaluated.

3. **Industrial equipment and machinery:** Evaluation of the operative yielding in each step of the manufacturing line analyzed.

4. **Material:** Evaluation of raw material used in the store and manufacturing areas, to obtain the final fabricated product, and analyze the necessary materials to make the industrial operations.

5. **Method:** Analysis of the way to make the industrial operations, standing or sitting, movements and times of the activities made in the new worktable.

6. **Measurement:** Evaluation of the appropriate measurements and compare them with quality standards established by specialized institutions or organizations.

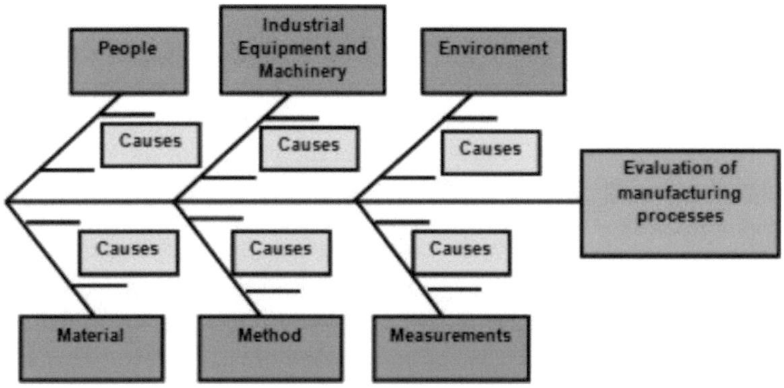

Figure 6.
Steps of Ishikawa diagram used in industries. Source: Analysis of the investigation.

2. Methodology

The study was very relevant to be able to establish a new process attached to the manufacturing line, and to determine with the COSIMIR software, the optimal conditions for the new industrial process. The investigation was elaborated in four phases, as explained below, explaining each of these:

a. Analysis with the Ishikawa diagram of times and movements of the operation of organization of microcomponents in the process in the conveyor belt.

b. Examination of modifying the process flow by eliminating a stage and being processed outside the conveyor belt.

c. Evaluation of friendly software for the design and simulation of industrial processes.

d. Analysis of the implementation in the manufacturing area of the electronic micro-industry.

3. Results

The implementation of the new work table once the design and simulation analyzes were made with the COSIMIR software determined the need to extract the operation of organizing biomechanical knee pads to avoid production stoppages and delays to the subsequent areas and to the customer with the final product to avoid production stopped and delays to the subsequent areas and to the customer with the final product. With the COSIMIR program, the work table was designed and manufactured and the required personnel were quickly trained. Once the expected results have been obtained, was proposed to replicate this procedure in other manufacturing line of this industrial company evaluated where the investigation was elaborated.

3.1 COSIMIR used in the industry

The initial focus of this software was for educational activities, subsequently using it for industrial operations research and the development of new prototypes. This software developed simulations to evaluate the possible improvements of adding structures, devices or working methods to obtain an immediate and lasting solution to problematic situations that was presented in the manufacturing areas. With this software, a new work table was designed that the SME biomedical industry did not have and was manufactured in this same company by design personnel of new processes and products. **Figure 7** shows the process of development of the work station used in this investigation and was designed by COSIMIR for the new industrial activity. The use of workstations in this industrial plant evaluated, increased in recent years, with the design of new prototypes for the development of manual activities, with the aim minimizing costs and maximizing economic gains, and that are ergonomically suited to operating personnel. The costs of these types of workstations sometimes exceed the expectations of companies that do not include such expenses in their budget, and due to this, they have been implemented in the manufacturing areas without optimizing operations in these.

3.2 Quality of ICT in industrial processes

In this investigation was evaluating the quality of the intensity of signals about the communication of data. This is showed in **Figure 8**, which is very important to determine the capacity of memory of the computer system in the manufacturing areas and the velocity to control industrial processes and the actual version of the COSIMIR software. As is observed in **Figure 8**, the different colors indicate the presence of diverse velocities of intensity of transition data by computers in the manufacturing areas. The development of a new workstation generated maximum efficiency originating an increase of levels of production (86%) and quality (89%) and decreased costs (82%) by having an optimization of processes based on the use of fingersine prepared in the COSIMIR software and similar evaluations that confirmed the need for this continuous improvement. This analysis was based in the reflection of less use of rework, less quantity of people in the manufacturing areas that were resintalled in a new manufacturing line of new product fabricated and the less quantity of fails and errors.

3.3 Correlation analysis of quality control

In this investigation, the distribution by flow of the product was evaluated, where, in this form of work, the distribution of the work tables was organized, as

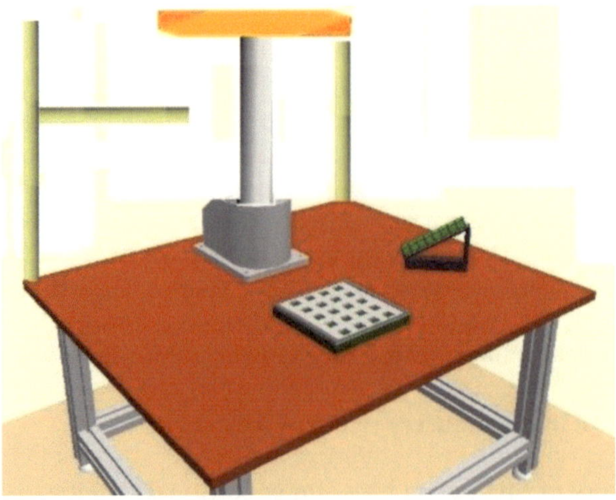

Figure 7.
Development process of the work table with the COSIMIR program.

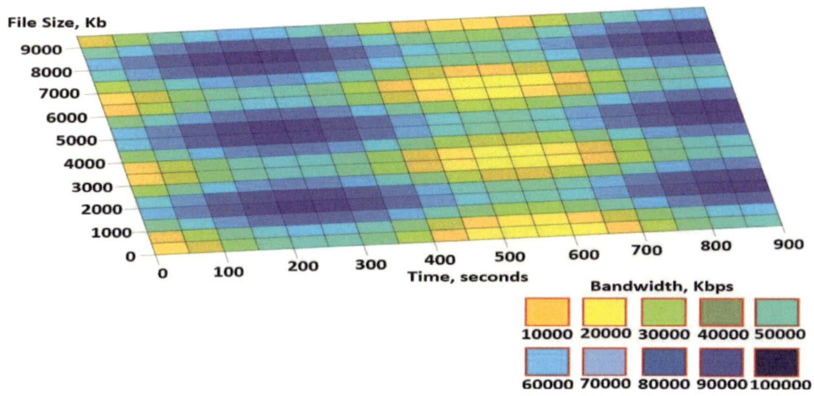

Figure 8.
Analysis of transition and reception of data in an computer system with COSIIMIR software (June, 2018).

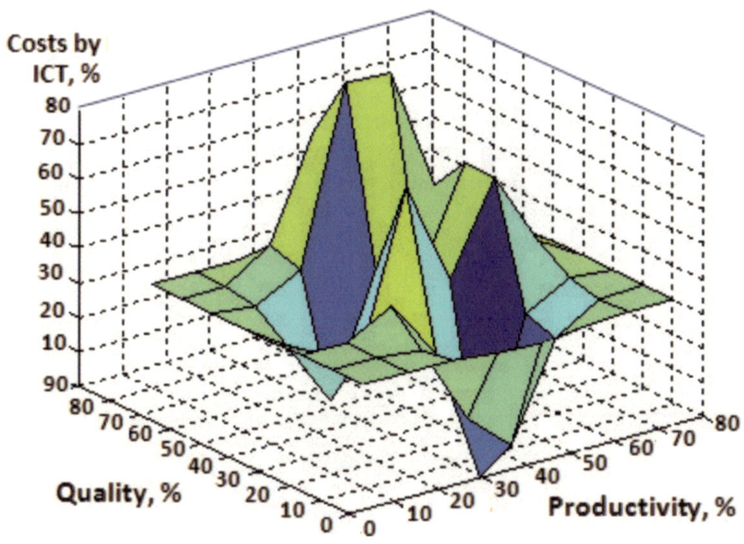

Figure 9.
Correlation analysis of productivity, quality and costs generated by ICT.

well as industrial equipment and machinery at each stage of the manufacturing process, with a sequence of operations to perform during the manufacturing of the product. The use of ICT has been considered as a new tool for business development, because transactions can be processed from anywhere in the world without having to wait for banking systems. This improves efficiency and management in the industries, which evaluate the costs of applying ICT in your company and carry out a comparative analysis of this aspect with the productivity and manufacturing quality indices. With this technological tool, what is called electronic commerce was formed, where financial operations are carried out quickly and easily from the matrixes of industries located in their countries of origin to where their branches are in other countries of the world. This is showed in **Figure 9**.

4. Conclusions

The use of ICT with the Cosimir software in the SME biomedical company, was very relevant in the evaluation of continuos improvement to determine the principal actions of the parameters evaluated of the operative yielding pf industrial equipment and machinery and the workers. With the simulations in the Cosimir software, in this new process and new worktable, was detected very easy and fastly, the functionability in the worktable and the possible fails and errors. This improve ensure the way to operation of the industrial process in the worktable, both to increase the production and quality indices and to eliminate errors, as well as to have operations of the operating personnel of this manufacturing area with the optimal conditions of postures and movements to avoid any health complications of the workers. The use of COSIMIR software was very relevant because specialized people of this company evaluated, can designa and fabricate a new workstation outside of the conveyor and improve the productivity and quality levels. Manufacturing processes were found to improve productivity and quality rates with automated electronic systems controlled by ICT. With the use of ICT such as COSIMIR, the evaluated micro-industry generated great reliability in its industrial processes and in its manufactured products, so that its profits increased.

Acknowledgements

The scientist are grateful for the support to the company where the investigation was made, which was elaborated with the economical and infrastructure of the participating industry and educational institution.

Author details

Roberto Carlos Valdés Hernández[1*], Juan Gabriel Lopez Hernandez[2], Adelaida Figueroa Villanueva[1] and Vidblain Amaro Ortega[3]

1 Facultad de Ciencias Administrativas, Universidad Autónoma de Baja California, Mexicali, Baja California México

2 Instituto de Ingenieria, Universidad Autonoma de Baja California, Mexicali, Baja California, Mexico

3 Departamento de Computación, Tecnológico Nacional de México, Instituto Tecnológico de Mexicali, Mexicali, Baja California, Mexico

*Address all correspondence to: carlos.valdes@uabc.edu.mx

References

[1] Grant Gabriel B, Seager Thomas P, Guillaume M, Loring N. Information and communication technology for industrial symbiosis. Journal of Industrial Ecology. 2010;**14**(5) Special Issue:Special Issue: Environmental Applications of Information & Communication Technology

[2] Alftan A, Kaipia R, Loikkanen L, Spens K. Centralized grocery supply chain planning: Improved exception management. International Journal of Physical Distribution and Logistics Management. 2015;**45**(3):237-259

[3] Carlos Valdés Hernández R, Arcos Vega JL, Fernando Navarro F, Flores Frias S. Client's satisfaction with software development quality in small and medium companies (PYMES) in Baja California, Mexico. International Journal of Computer Science and Engineering (IJCSE). 2017;**6**(6):1-8

[4] Carrasco Martínez AC. 2010. Estudio ergonómico en la estación de trabajo PT0780 de la empresa S-Mex, S.A de C.V. León. México. Tesis de Ingeniería Industrial. Huajuapan de Leóm, Oaxaca: Universidad Tecnológica de la Mixteca. 2010

[5] Dymora P, Koryl M, Mazurek M. Process discovery in business process management optimization. Information. 2019;**10**(9):270

[6] Valdez H, Roberto C. 2019. Caracterización del Desarrollo, Satisfacción del Cliente y Calidad del Software en Pymes de Baja California, México. Tesis de Doctorado. Baja California, Mexico: Universidad Autonoma de Baja California-Mexicali. p. 124

[7] INDEX-Baja California. 2019. Reporte de empresas maquiladoras de Baja California. INDEX-BC

[8] Menéndez V, Castellanos M. SPEM: Software process engineering metamodel. Revista Latinoamericana de Ingenieria de Software. 2015;**3**(2):92-100

[9] Henderson F, Young H. FESTO. Educational User Guide of COSIMIR software used in manufacturing processes. Ed. FESTO Company; 2003. pp. 1-78

[10] Benavides D, Felfernig A, Galindo JA, Reinfrank F. Automated analysis in feature modelling and product configuration. In: International Conference on Software Reuse. Berlin, Heidelberg: Springer; 2013. pp. 160-175

[11] Walas Mateo F, Lastiri V, Figari Bizzoto S, Andrieu D. Estudio del impacto de implementación de tecnología en la competitividad de Las cadenas de valor sectoriales en la Región de influencia de la UNAJ. Revista ED Experiencia Docente. 2014;**2**:1

[12] Garcia A, Trujillo Y, Perdomo A. Optimización de estados en la mejora de procesos de software. Enlace: Revista Venezolana de Información, Tecnología y Conocimiento. 2016;**13**(2):9-27

[13] Gavin M, Siedschlag. Human capital and growth of information and communication technology-intensive industries: Empirical evidence from open economies. Regional Studies Journal. 2013;**47**(9):1403-1424. DOI: 10.1080/00343404.2010.529115

[14] Kalpakjian S. Manufacturing engineering and technology. 2nd. Ed. Editorial Prentice Hall Ed.; 2014. p. 89

[15] Magrab E, Azarm S, Balachandran B, Duncan J, Herold K, Walsh G. An Engineer's Guide to MATLAB, 3e: With Applications from Mechanical, Aerospace, Electrical,

and Civil Engineering. Editortial Prentice Hall; 2011. p. 846. ISBN: 978-0-13-1991190-1

[16] Groover M. Fundamentals of modern manufacturing: Materials, processes, and systems. 3rd. Ed. Editotial Wiley; 2007. p. 121

[17] Potnuru D, Alice Mary K, Ch S. Designandimplementationmethodology for rapid control prototyping of closed loop speed control for BLDC motor. Journal of Electrical Systems and Information Technology. 2018;5(1):99-111

Chapter 11

Fuzzy Monte Carlo Simulation to Optimize Resource Planning and Operations

Mohammad Ammar Alzarrad

Abstract

Resources planning and operations are essential concerns and specialty areas within industrial engineering and project management. Crew configuration plays a significant role in resource planning and operations. Crew configuration inefficiency is one of the most common reasons for the low productivity of manpower. Resources planning contains some inherent uncertainties and risks because it is an estimate of unknown values. Many factors affect resource planning. Some of these factors are fuzzy variables such as expert's judgment, and some of them are random variables such as direct cost of equipment. The objective of this chapter is to present a method that combines fuzzy logic and Monte Carlo simulation (MCS) for the selection of the best crew configuration to perform a certain task. The model presented in this chapter is a joint propagation method based on both the probability theory of MCS and the possibility theory of fuzzy arithmetic. The research outcomes indicate that the presented model can reduce the duration and cost of a certain task, which will help reduce the cost and duration of the project.

Keywords: fuzzy logic, Monte Carlo simulation, manpower productivity, resources planning, optimization

1. Introduction

Project schedule can often be shortened by assigning excess crews (labor and equipment) to critical activities. However, the ultimate cost consequences to the project are often difficult to estimate. This schedule compression strategy may adversely impact project cost performance because the overstaffing of critical activities may result in wasted or idle time in these activities. Trade-offs between elapsed time and the associated cost of crew is required to determine the best crew configurations [1]. This is not a trivial issue because of the complex relationships between elapsed time, crew configurations, and their associated costs.

Monte Carlo simulation (MCS) has been used widely to solve probabilistic uncertainty in range estimating for projects [2]. It has been extensively used for generating many scenarios by considering the random sampling of each probability distribution. In practice, the probability of an event can be estimated according to the frequency of that event occurring in a number of experiments [3]. However, if the number of experiments is not large enough to be significant, and more experiments cannot be performed, it is not possible to accurately estimate the event's probability. In these circumstances, we can engage human experts who are usually good at supplying the required information. Some researchers try to convert experts' knowledge into probabilistic distributions. However, this can lead to pointless and unreliable results since the results are obtained based on experts' subjective judgments and assumptions [4]. Fuzzy Logic has been used successfully for representing such uncertainties in experts' judgments [5].

This research proposes a fuzzy Monte Carlo simulation (FMCS) model that provides the capability of considering fuzzy and probabilistic uncertainty simultaneously to help improve decisions regarding crew configurations.

2. Monte Carlo simulation

Monte Carlo simulation (MCS), or probability simulation, is a technique used to understand the impact of risk and uncertainty cost, time, and other forecasting models [4]. MCS estimates the expected value based on historical data, or expertise in the field, or experience. While this estimate is useful for creating a model, it contains some intrinsic uncertainties, because it is an estimate of unknown values [4].

In project management, you could use expert knowledge to estimate the time it will take to complete a particular job, you can also estimate the maximum time it might take, in the worst possible case, and the minimum time, in the best possible case. The same could be done for project costs. The Monte Carlo simulation method is used for estimating the output Y of a function (M) with random input variables (R1, R2, ... , Rn) (**Figure 1**).

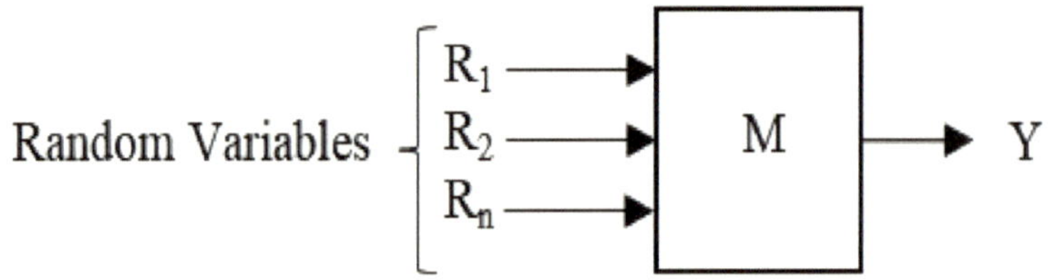

Figure 1.
The output Y of a function M with random inputs can be calculated using Monte Carlo simulation.

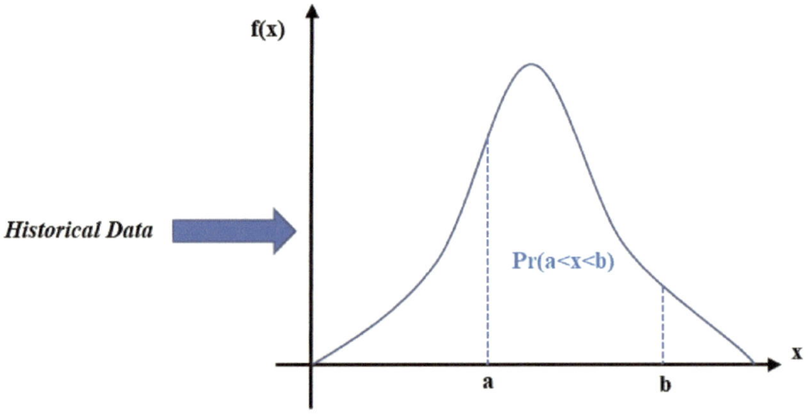

Figure 2.
A probability density function (PDF) developed based on historical data.

In a Monte Carlo simulation, an arbitrary value is selected for each of the activities, based on the range of estimates. The model is calculated based on this arbitrary value. The result of the model is recorded, and the process is repeated [6]. A traditional Monte Carlo simulation calculates the model hundreds/thousands of times, each time using different randomly selected values. When the simulation is complete, we have a large number of outcomes, each based on random input values. These outcomes are used to describe the likelihood, or probability, of reaching various results in the model [6] (**Figure 2**).

3. Fuzzy logic

Fuzzy logic is a technique that offers a clear conclusion from unclear and inaccurate data. The Fuzzy Set concept was first introduced by Zadeh in 1965 [7]. He was inspired by witnessing that human thinking could utilize ideas that do not have precise borders [8]. Fuzzy logic and fuzzy hybrid methods have been used to capture and model risk, thereby improving workforce and project management [8]. Fuzzy logic can effectively capture expert knowledge and engineering judgment and combine these subjective elements with project data to improve construction decision making, performance, and productivity [9]. The triangular fuzzy number (TFN) is a common shape of fuzzy logic (**Figure 3**). The α-cut method is a common technique to do arithmetic operations on a Triangular Membership Function [10]. The α-cut signifies the degree of risk that the project managers are ready to take (i.e., no risk to full risk). Because the value of α could significantly affect the solution, it should be wisely chosen by project managers [11].

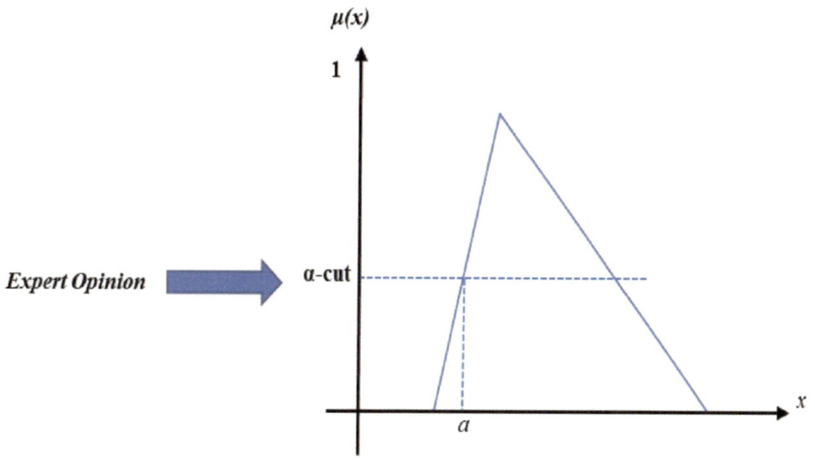

Figure 3.
Triangular fuzzy set developed based on experts' judgment.

4. Fuzzy Monte Carlo simulation (FMCS)

The proposed FMCS is a joint propagation method based on both the probability theory of MCS and the possibility theory of fuzzy arithmetic. A generalized problem in which we have both types of uncertainty, fuzzy and probabilistic. Here, we need to determine the output Y of a function (M) that has R_1, R_2, … , R_n being random

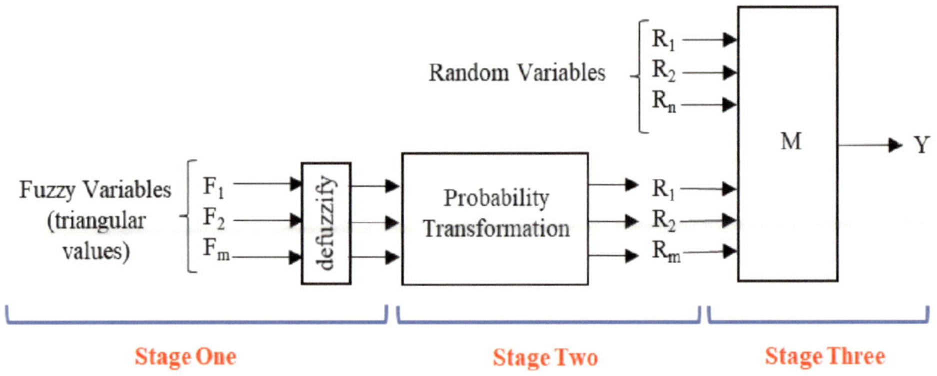

Figure 4.
Converting fuzzy sets to PDF before performing Monte Carlo simulation.

variables and represented by probabilistic distributions and F_1 (triangular values), F_2 (triangular values), ..., F_m being fuzzy sets (**Figure 4**).

The first stage of the model is to defuzzify the fuzzy variables to get crisp values. The centroid method is one of the most common methods for defuzzification, in which the defuzzified value is calculated by finding the center of the area under the membership function.

The second stage of the model is to transfer the crisp sets of numbers that we got in step one to random variables. To that, the confidence level of the intervals is estimated using the probability of that interval. This probability is equal to the area under the PDF that is bounded within that interval. Among different intervals of the same confidence level, the most informative interval is the one with minimal length [8].

The third stage of the model is to use an optimization algorithm for finding a set of optimal values minimizing or maximizing the given object function [MIN F(X) or MAX F(X)] subjected to minimum-maximum intervals. Then, the uncertainty quantifications propagated using the obtained optimal values are represented as a plausibility distribution and a belief distribution. **Figure 5** shows Fuzzy Monte Carlo simulation (FMCS) process.

5. Case study

To illustrate an implementation of the FMCS model, the researcher analyzes the behavior of FMCS framework in comparison with traditional Monte Carlo simulation using a time range estimating example. Consider a sample application by [12] of a time range estimating problem for an excavation project. The time and cost needed for the project equipment are shown in **Table 1**, and probabilistic distributions are used to express the uncertainty regarding those variables.

These uncertainties may result from uncertainty regarding different scenarios that may happen in the field during construction. For example, uncertainty in the activity duration may be a result of uncertainty associated with the productivity of workers or variability in weather conditions.

The case study evaluated seven alternatives using Monte Carlo simulation (MCS), to compare the presented FMCS model with the traditional MCS that is used by [12], we will evaluate the same alternatives to see if the outcomes of the new FMCS model will be different. The seven equipment configurations alternatives are shown in **Table 2**.

Using Fuzzy logic toolbox in MATLAB the FMCS generates the results shown in **Table 3**.

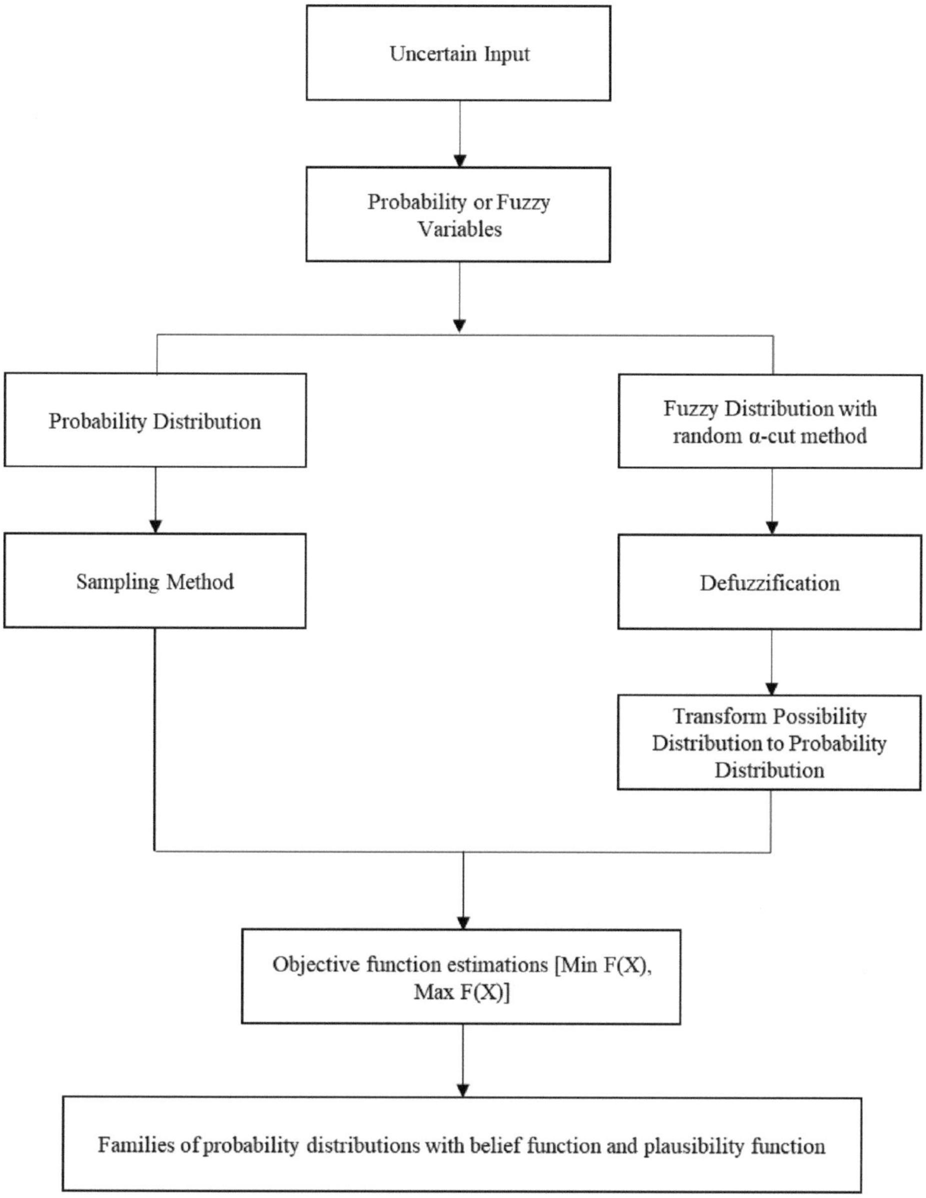

Figure 5.
Fuzzy Monte Carlo simulation (FMCS) process.

Activity	Triangular duration (min)	Triangular cost ($/min)
20 cu yd hauler: haul, unload, return	(18,33,48)	2.39
15 cu yd hauler: haul, unload, return	(15,28,41)	2.29

Table 1.
Activity time and cost data.

From **Table 3**, we can conclude that the best alternative is number seven with shift duration of 17 days and shift cost of $79,050. By comparing this result with the result published by [12], we find out that [12] concluded that the best alternative is also alternative seven with shift duration of 18 days and shift cost of $82,665.

Alternatives	Configurations
Alt 1	4 of 20-cu yd haulers
	2 of 15-cu yd haulers
Alt 2	5 of 20-cu yd haulers
	2 of 15-cu yd haulers
Alt 3	4 of 20-cu yd haulers
Alt 4	3 of 20-cu yd haulers
Alt 5	2 of 20-cu yd haulers
	2 of 15-cu yd haulers
Alt 6	8 of 20-cu yd haulers
	3 of 15-cu yd haulers
Alt 7	6 of 20-cu yd haulers
	3 of 15-cu yd haulers

Table 2.
Alternatives of equipment configurations.

Alternatives	Shift duration (days)	Shift cost ($)
Alt 1	26	77,438
Alt 2	26	86,667
Alt 3	27	63,780
Alt 4	31	67,549
Alt 5	33	61,170
Alt 6	17	98,014
Alt 7	**17**	**79,050**

Table 3.
FMCS results.

6. Conclusion

This chapter proposes a new FMCS model. This model can consider both fuzzy and probabilistic uncertainty in the resource planning problem. The model provides decision-makers with the ability to reduce risk in the output in the form of fuzziness and probabilistic uncertainty, and he/she can use subjective judgment and experience to make the final decision. A case study has been analyzed to evaluate the robustness of the FMCS model. The outcomes show that the proposed FMCS model could help reduce the duration and cost of a certain task, which will help reduce the cost and duration of the project.

Acknowledgements

The research leading to the publication of this chapter was partially supported by the Department of Civil Engineering at Marshall University in Huntington, West Virginia, United States of America.

Conflict of interest

The authors declare that there are no conflicts of interest regarding the publication of this chapter.

Author details

Mohammad Ammar Alzarrad
Marshall University, Huntington, USA

*Address all correspondence to: alzarrad@marshall.edu

References

[1] Alzarrad MA, Fonseca D. A new model to optimize project time-cost trade-off in an uncertain environment. In: Moynihan G, editor. Contemporary Issues and Research in Operations Management. Rijeka, Croatia: IntechOpen; 2018. pp. 95-112. DOI: 10.5772/intechopen.74022

[2] Morin MA, Ficarazzo F. Monte Carlo simulation as a tool to predict blasting fragmentation based on the Kuz-Ram model. Computers & Geosciences. 2006;**32**(3):352-359

[3] Kim Y. Monte Carlo vs. fuzzy Monte Carlo simulation for uncertainty and global sensitivity analysis. Sustainability. 2017;**9**(4):539-554. DOI: 10.3390/su9040539

[4] Mitropoulos LK, Prevedouros PD, Yu X, Nathanail EG. A fuzzy and a Monte Carlo simulation approach to assess sustainability and rank vehicles in urban environment. Transportation Research Procedia. 2017;**24**:296-303. DOI: 10.1016/j.trpro.2017.05.121

[5] Alzarrad MA, Moynihan GP, Hatamleh MT, Song S. Fuzzy multi-criteria decision-making model for time-cost-risk trade-off optimization in construction projects. Advances in Civil Engineering. 2019;**4**:1-7. Article ID: 7852301. DOI: 10.1155/2019/7852301

[6] Sadeghi N, Fayek AR, Pedrycz W. Fuzzy Monte Carlo simulation and risk assessment in construction. Computer-Aided Civil and Infrastructure Engineering. 2010;**25**(4):238-252. DOI: 10.1111/j.1467-8667.2009.00632.x

[7] Zadeh LA. Fuzzy logic and approximate reasoning. Synthese. 1975; **30**(34):407-428

[8] Li J, Moselhi O, Alkass S. Forecasting project status by using fuzzy logic. Journal of Construction Engineering and Management. 2006;**132**(11):1193-1202

[9] Fayek AR, Jose RRF. Application of fuzzy logic to quality assessment of infrastructure projects at conceptual cost estimating stage. Canadian Journal of Civil Engineering. 2010;**37**(8): 1137-1147

[10] Mohagheghi V, Mousavi SM, Vahdani B. Analyzing project cash flow by a new interval type-2 fuzzy model with an application to construction industry. Neural Computing and Applications. 2017;**28**(4):3393-3411

[11] Ayhan MB. Fuzzy TOPSIS application for supplier selection problem. International Journal of Information Management. 2013;**5**(2): 159-174

[12] Maxwell D, Back W, Toon J. Optimization of crew configurations using activity-based costing. Journal of Construction Engineering Management. 1988;**124**(2):162-168

Chapter 12

Weapon Target Assignment

Mohammad Babul Hasan and Yaindrila Barua

Abstract

This chapter is mainly based on an important sector of operation research-weapon's assignment (WTA) problem which is a well-known application of optimization techniques. While we discuss about WTA, we need some common terms to be discussed first. In this section, we first introduce WTA problem and then we present some prerequisites such as optimization model, its classification, LP, NLP, SP and their classifications, and applications of SP. We also discuss some relevant software tools we use to optimize the problems. The weapon target assignment problem (WTA) is a class of combinatorial optimization problems present in the fields of optimization and operations research. It consists of finding an optimal assignment of a set of weapons of various types to a set of targets in order to maximize the total expected damage done to the opponent. The WTA problem can be formulated as a nonlinear integer programming problem and is known to be NP-complete. There are constraints on weapons available of various types and on the minimum number of weapons by type to be assigned to various targets. The constraints are linear, and the objective function is nonlinear. The objective function is formulated in terms of probability of damage of various targets weighted by their military value.

Keywords: weapon, assignment, transportation, non-linear programming

1. Introduction

This Chapter is mainly based on an important sector of operation research-weapon's target assignment (WTA) problem which is a well-known application of optimization techniques. While we discuss about WTA, we need some common terms to be discussed first. In this section, we first introduce WTA problem and then we present some prerequisites such as optimization model, its classification, LP, NLP, SP and their classifications, and applications of SP. We also discuss some relevant software tools we use to optimize the problems.

1.1 Weapon target assignment

The weapon target assignment problem (WTA) is a class of combinatorial optimization problems present in the fields of optimization and operations research. It consists of finding an optimal assignment of a set of weapons of various types to a set of targets in order to maximize the total expected damage done to the opponent. The WTA problem can be formulated as a nonlinear integer programming problem and is known to be NP-complete. There are constraints on weapons available of various types and on the minimum number of weapons by type to be assigned to various targets. The constraints are linear, and the objective function is nonlinear. The objective function is formulated in terms of probability of damage of various targets weighted by their military value.

IntechOpen

1.2 Preliminaries

In the current section, we discuss some preliminaries of the terms we mention in the chapter.

1.2.1 Optimization model

Optimization means 'the action of finding the best solution'. Optimization modeling is also known as Mathematical Programming. Mathematical programming is the use of mathematical models, particularly optimizing models, to assist in making decisions. It is a branch of operation research which has wide applications in various areas of human activity. Optimization can help solve problems where there are two situations as (1) many ways of doing something or (2) limited resource available.

1.3 Classification of optimization problem

Any real-world optimization problem may be characterized by five qualities. The problem function may all be linear or be nonlinear. The functional relationships may be known i.e. deterministic, or there may be uncertainty about them i.e. probabilistic. The optimization may take place at a fixed point in time (static) or it may be an optimization over time (dynamic). The variables may be continuous or discrete. And lastly, the problem functions may all be continuously differentiable (smooth) or may have points where the functions are non-differentiable (non-smooth).

1.4 Linear programming (LP)

Linear programming is an optimization technique of a linear objective function, subject to linear equality and linear inequality constraints. It is a mathematical method that is used to determine the best possible outcome or solution from a given set of parameters or a list of requirements, which are represented in the form of linear relationships. It is most often used in computer modeling or simulation in order to find the best solution in allocating finite resources such as money, energy, manpower, machine resources, time, space and many other variables. In most cases, the "best outcome" needed from linear programming is maximum profit or lowest cost. It was first developed by Soviet mathematician and economist Leonid Kantorvich in 1937 during the second world-war.

1.5 Standard form of LP

Here we present the standard form of linear programming. A linear programming problem may be defined as the problem of maximizing or minimizing.

The standard linear programming problem can be expressed in a compact form as:
Maximize (or Minimize)

$$z = \sum_{i=1}^{n} c_i x_i \tag{1}$$

$$\text{subject to } \sum\nolimits_{j=1}^{m} a_{ij} x_j \{\leq, =, \geq\} b_i, \quad i = 1, 2, \dots, n \tag{2}$$

$$x_j \geq 0 \quad j = 1, 2, \dots, m$$

The basic components of linear programming are as follows:

- Decision variables (x_j)—These are the quantities to be determined.
- The objective function (1)—This represents how each decision variable would affect the cost, or, simply, the value that needs to be optimized.
- Constraints (2)—These represent how each decision variable would use limited amounts of resources.
- Data—These quantify the relationships between the objective function and the constraints. c_i is called profit or cost coefficients, a_{ij} are the constraint coefficients and b_i are the availability of resources or minimum requirement.

1.6 Stochastic programming (SP)

The aim of stochastic programming is to find optimal decisions in problems which involve uncertain data. For optimization under uncertainty stochastic programming is one of the best techniques. That is, stochastic programming is mathematical programs that include data that is not known with certainty but is approximated by probability distributions. Stochastic programming extends the scope of linear and nonlinear programming to include probabilistic or statistical information about one or more uncertain problem parameters. Similarly, when all the input data used in the mathematical formulation of the mathematical program is known with certainty then the corresponding models are called deterministic models.

1.7 Types of SP problem

Stochastic programming offers a solution by eliminating uncertainty and characterizing it using probability distributions. There exist many different types of stochastic problems. The most famous type of stochastic programming model is recourse problems. Another form of a stochastic problem is the chance-constrained programming problem. In this type of stochastic programming model, the constraints to be optimized depend on probabilities. The classification of SP problems is shown in **Figure 1**.

1.8 Applications of stochastic programming

Stochastic programming has been applied to a wide variety of areas. Some of the specific problems are part of the Stochastic Programming test set. Other applications are listed as follows: Manufacturing Production Planning, Manufacturing production capacity planning, Machine Scheduling, Freight scheduling, Dairy Farm Expansion planning, Macroeconomic modeling and planning, Timber management, Asset Liability Management, Portfolio selection, Traffic management, Optimal truss design, Automobile Dealership inventory management, Lake level management.

1.9 Software tools

Nowadays computerized techniques are widely used to solve various types of problems in the world. Sometimes some problems become difficult to solve and time-consuming by hand calculation. So by using different software tools, we can

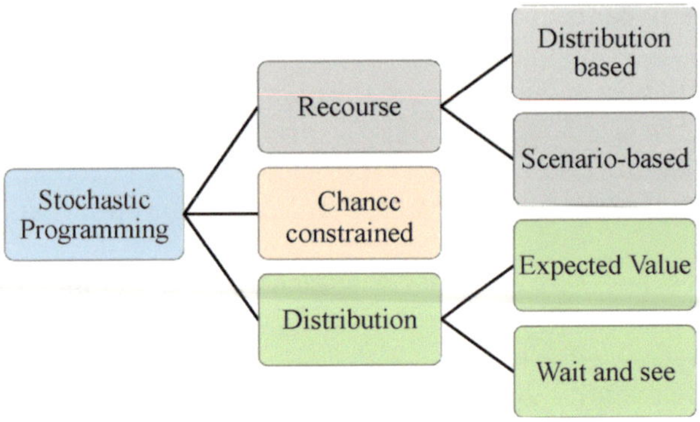

Figure 1.
Codification of SP problems.

solve problems from small to large scale problem optimally in a short time. There are so many computer-based mathematical programming languages have been used worldwide. Some of the tools that are used to solve optimization problems are **LINDO, LINGO, AMPL, MATLAB, MATHEMATICA, MAPLE, MS EXCEL SOLVER** and **TORA,** etc. In this chapter, we use AMPL and LINGO.

1.9.1 AMPL

AMPL, an acronym for "A Mathematical Programming Language" is a comprehensive and powerful algebraic modeling language for linear and nonlinear optimization problems, with discrete or continuous variables. It is a language for solving high complexity problems for large scale mathematical computation. It was developed by Robert Fourier, David Gay and Brian Kernighan at Bell Laboratories [1]. By using AMPL, we can get the solution of a problem in which the model of the formulation with sets, variables, parameters, constraints, etc. are written in a mod. file and the data of the formulation are written in a dat. file. Then the solution is found after running the program in the console window.

1.9.2 LINGO

LINGO is designed to solve a wide range of optimization problems, including linear programs, mixed integer programs, quadratic programs, stochastic, and general nonlinear non-convex programs faster, easier and more efficient. It provides a completely integrated package that includes a powerful language for expressing optimization models, a full-featured environment for building and editing problems, and a set of fast built-in solvers.

1.10 Motivation

First one is based on weapons assignment in which the engagement of a target by a weapon is modeled as a stochastic event. In this type of problem, we develop a general computer oriented algorithm so that we can solve this type of problems for small scales to large scales problems in a single framework. To show the effectiveness of our developed model we present numerical examples of WTAP and compare our result with different existing results.

1.11 Outline of the chapter

This chapter contains four sections in total which is organized as follows:

- In Section 1, we discuss some prerequisites that are required for WTA problem. We also discuss about the types of optimization models, software tools that we use in this chapter.

- In Section 2, we review some relevant papers about weapon's assignment problem.

- In Section 3, we discuss the weapon target assignment problem. We formulate the WTA problem. Some existing algorithms are also presented in this chapter. We discuss the real-life applications and present numerical examples of WTAP. We develop a new computer technique by using programming language AMPL to solve all type of WTA problem in a single framework. Then finally compare the results we get from AMPL to previously solved results of the examples.

- In Section 4, we draw a conclusion about our whole chapter.

1.12 Conclusion

In this Section, we discussed the relevant preliminaries. In the next Section, we will review some literature about weapons assignment and chance-constrained problem.

In this section, we will review some admissible research articles on Weapon Target Assignment Problem. Since the 1950s, the optimal assignment problem of weapons to targets has always been concerned by many countries. The study of WTA problem can be traced back to the 1950s and 1960s when Manne [2] and Day [3] built the model of WTA problem. The present research work on WTA is focused on models and algorithms. In the research on models of WTA, the static WTA

Researchers	Year	Proposed Algorithms	Implementation WTA
Galat and Simaan	2007	Tabu	Dynamic single-objective
Lee	2010	VLSN	Static single-objective
Xin et al.	2010	VP + Tabu	Dynamic single-objective
Li and Dong	2010	DPSO+SA	Dynamic single-objective
Chen et al.	2010	SA	Static single-objective
Fei et al.	2012	Auction Algorithm	Static single-objective
Liu et al.	2013	MOPSO	Static multi-objective
Zhang et al.	2014	MOEA/D	Static multi-objective
Ahner and Parson	2015	Dynamic Programming	Dynamic multi-objective
Li et al.	2015	NSGA-II, MOEA/D	Static multi-objective
Driik et al.	2015	MILP	Dynamic multi-objective
Liang and Kang	2016	CSA	Static single-objective
Li et al.	2016	MDE	Dynamic multi-objective

Table 1.
Existing algorithms for several ye.

models are mainly studied and the dynamic WTA are not fully studied indeed. In the research on algorithms of WTA, the intelligent algorithms are often used to solve the WTA problem.

There are so many proposed algorithms on WTA problem [4, 5]. So we present the summary of variant heuristic algorithms and the implementation of various WTA have been proposed for several years is shown in **Table 1**.

2. Weapon's target assignment

Various combinatorial optimization techniques are currently available. Most of these techniques have not been thoroughly tested on realistic problems. In this chapter, we consider a class of non-linear assignment problems collectively referred to as Target-based Weapon Target Assignment (WTA). We first briefly discuss the weapon target assignment problem. We also include the basic concepts and models of WTA and the mathematical nature of the WTA models is also analyzed. We present some real-life applications of WTA here. There does not exist any exact methods for the WTA problem even relatively small size problems, and much research has focused on developing heuristic algorithms based on meta-heuristic techniques. The main focus of this chapter is our new developed optimization algorithm for the WTA problem based on the kill probabilities.

3. Weapon target assignment (WTA) problem

The assignment problem is one of the fundamental constrained combinatorial optimization problems in the branch of optimization or operation research in Mathematics. This problem is mainly used in decision making. Here we consider a special type of problem which is a combination of transportation problem and assignment problem. By the name of weapon-target assignment problem, it is clear that we have to assign weapons to targets. It is a defense-related application in operation research and is slightly different from the more general optimal resource allocation problem. The main aim of weapon-target assignment problem is to find a set of solution of the number of available weapons to a set of required targets so that the expected rewards of the sequential engagement is maximized [6]. The engagement of a target by a weapon is modeled as a stochastic event, with a probability of kill assigned to each weapon-target pair (this is the probability that the interceptor weapon will destroy the target if assigned to it). The engagement of a weapon-target pair is independent of all other weapons and targets. This is an integer optimization problem in that fractional weapon assignments are not allowed.

3.1 Basic factors of WTA

A number of different approaches have been applied to the WTA problem. When considering a WTA problem, a number of factors need to be considered. Some of these factors are discussed below [7]:

3.1.1 Linear versus non-linear assignment problem

The generalized linear assignment problem (LAP) of allocating weapons to targets is a fundamental problem of combinatorial optimization. In the simplest case, the number of weapons and targets are equal, with only one weapon being assigned to any one target in an allocation. LAP's can also be represented in a bipartite graph shown in **Figure 2 (a)**. In the LAP graph, weapons cannot be

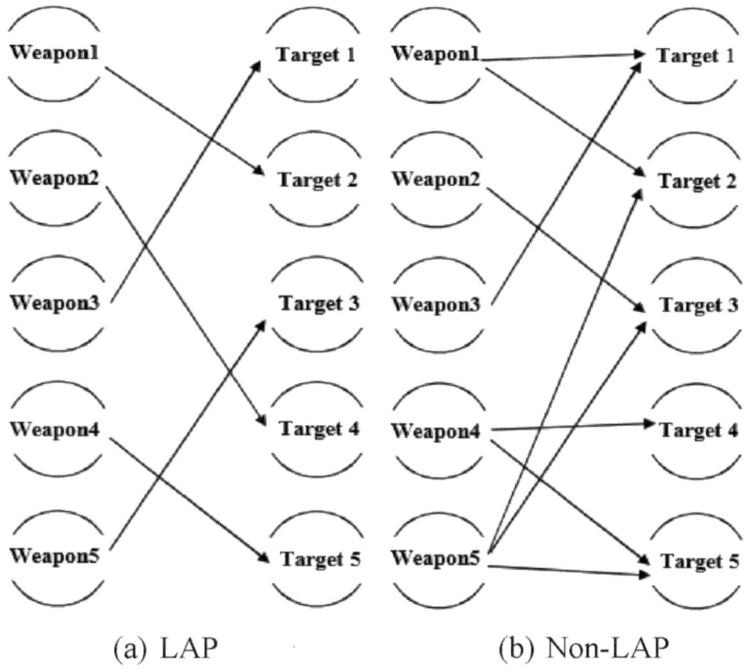

Figure 2.
A linear and a nonlinear bipartite graph.

assigned to more than one target. But, when targets are assigned to more than one targets or targets, are assigned to more one weapon, then the assignment problem becomes nonlinear as presented by the bipartite graph in **Figure 2 (b)**.

Weapon target assignments are generally viewed as nonlinear assignment problems (non-LAP). That is, the optimal solution is nonlinear but is still considered to integer values as in the LAP case.

3.1.2 Asset-based versus Target-based

A WTA problem can be viewed from either a target-based or an asset-based perspective. In the target-based, values are assigned to each target to cause damage to the defended asset. The objective of the target-based WTA solution is to maximize the damage value of the incoming targets.

Conversely, in an asset-based perspective values are assigned to the assets rather than the targets. This WTA problem is where weapons are assigned such that the combined value of assets is maximized. The asset-based approach requires information on which targets are approaching the defended assets. But a target-based approach is more appropriate than the asset-based. The approach which is discussed in this chapter is the target-based perspective.

3.1.3 Static versus dynamic

Generally, WTA is categorized into two versions:

a. Static WTA

b. Dynamic WTA

Static WTA: In the static version, all of the inputs (i.e., weapons, targets, desired effects, engagement time, etc.) to the problem are fixed, and all weapons are engaged to targets in a single stage. The damage assessment is made when all the weapons are engaged to the targets completely. Thus the main objective of SWTA [8] is to find the proper assignment of a temporary defense task. That is, in static WTA the optimal assignment of weapons to targets only allowed a single weapon to be assigned to a single target. Then the static one can be considered as a constrained resource assignment problem. The static version of the problem is a special case of the dynamic one.

Dynamic WTA: Dynamic WTA problem is originally proposed by Hosein and Athans in 1990 [9], and attract much more attention from researchers in recent years. The goal of DWTA is to find a global optimal assignment for the whole defense process in which the engagement occasion of weapons must be taken into account. The dynamic WTA can also be expressed as a succession of static WTA. That is, in dynamic WTA there are no restrictions as discussed before in SWTA problem. Many weapons can be assigned to a single target. This satisfies the real scenario of a defense system. When the scale is large, the DWTA models are comparatively more complex than the SWTA models. In this paper, we mainly focus on the dynamic weapon-target assignment problem.

In addition, considering the different missions, each version includes the asset-based problem and the target-based problem. In the asset-based problem, the task is to maximize the expected total value of assets which are defended by the defensive weapons. In the target-based problem, the task is to minimize the expected total value of targets which are not destroyed by the defensive weapons after the engagement. The target-based problem can be considered as a special case of the asset-based problem.

3.1.4 Properties of dynamic WTA

Some relevant properties of the dynamic WTA problem are that it is:

a. NP-Complete (Non-deterministic polynomial), that is one must essentially resort to complete enumeration to find the optimal solution.

b. Discrete (fractional weapons assignment are not allowed)

c. Dynamic (the results of previous engagements are observed before making present assignments)

d. Nonlinear (the objective function is convex)

e. Stochastic (weapon-target engagements are modeled as stochastic events)

f. Large-Scale (the number of weapons and targets is large, making enumeration techniques impractical).

These properties of the problem rule out any hope of obtaining efficient optimal algorithms.

3.2 Mathematical formulation of WTA

To present the dynamic weapon-target assignment problem, we need the following parameters and variables to be introduced:

Symbols: Descriptions.

W: The number of Weapon types.

T: The number of targets that must be assigned by weapons.

u_j: The military value of the target js. This is determined during the weapon assignment and used to priorities target engagement.

w_i: The number of weapons of type i available to be assigned to targets.

t_j: The minimum number of weapons required to target j.

$p^{ij}{}_{kill}$: The destroying probability of target j by weapon of type$i$, also expressed as the kill probability for weapon type i on target j. It's given for all weapons and targets.

x_{ij}: An integer decision variable indicating the number of weapons of type i assigned to target

j: That is, how many numbers of weapons of type i should be assigned to target j to maximize the expected damage value of targets.

Let there be T targets numbered as $1, 2, \dots, T$ and W weapon types numbered $1, 2, \dots, W$. Then we can now formulate the objective function in terms of probability of damage of various targets weighted by their military value. So the weapon target assignment may now be modeled as the following nonlinear integer programming formulation in terms of the above-introduced variables,

$$\max \quad \sum_{j=1}^{T} u_j \left[1 - \prod_{i=1}^{W} \left(1 - p^{ij}{}_{kill}\right)^{x_{ij}} \right], \tag{3}$$

$$\textit{subject to,} \quad \sum_{j=1}^{T} x_{ij} \le w_i \qquad i = 1, \dots, W \tag{4}$$

$$\sum_{i=1}^{W} x_{ij} \ge t_j \qquad j = 1, \dots, T \tag{5}$$

$$x_{ij} \ge 0, \text{integer}, \; i = 1, \dots, W \;\; \text{and}\, j = 1, \dots, T \tag{6}$$

Here we consider $p^{ij}{}_{kill}$ as the destroying probability by weapons of type i on target j, therefore the term $\left(1 - p^{ij}{}_{kill}\right)$ denotes the survival probability for target j if weapon i assigned to it. By the over-all assignment of weapons of all types, $\sum_{i=1}^{W} x_{ij}$ the expected damage to target j is $\left[1 - \prod_{i=1}^{W} \left(1 - p^{ij}{}_{kill}\right)^{x_{ij}}\right]$. The maximization of the probability of the total expected damaged value of the targets is being represented by the objective function (3). Limitations on the number of weapons assigned are specified in terms of w_i and t_j. The constraints represented above by Eqs. (4) and (5) are on weapons available of various types and on the minimum number of weapons to be assigned to various targets. By the Eq. (4), we can assure that the total number of weapons used does not exceed the available capacity, and as well as the Eq. (5) ensures that the total number of weapons should exceed the minimum number of weapons required for target$j$s. Eq. (6) provides the non-negativity of decision variables. Here we observe that the resulting problem has non-linear objective function and linear constraints.

3.2.1 Applications of WTA

The WTA problem has wide applications in real life. Some of them are discussed in the current section:

3.3 Air missile system

In an air missile defense system [10, 11], missiles are regarded as the major weapon in modern warfare, and missile defense technology becomes a hot research topic for military and information expert. The reasonable target assignment strategy and optimization algorithm for weapon-target assignment improve operational effectiveness greatly. According to target threat degree and air combat priority index of target intercepted, the relative weigh for weapon unit of target attack is definite, the combined effect on target assignment result is weighed, which ensure high target interception as far as possible. In multi-fighter air combat, the weapon target assignment problem is a challenge in information warfare, the air defense command system can assign weapon reasonably for eliminating the threat from enemy targets in time. The selection rules of target function include the facts such as less resource and energy loss for fighter, the minimum threat degree and the minimum number of targets remaining, different selection rule reflect different decision intention, which decided different target function form and combat strategy [12]. AS an NP-complete problem, with the number increasing in weapon units and targets, the solution space shows the trend of the combined explosion [13].

3.4 WTA approach to media allocation

In management science, the word advertisement is the most significant term. In advertising, media allocation is a very important task for advertisers. Communication vehicles such as television, newspapers, internet, radio and etc. are referred by the term media in advertising. To convey the commercial messages to target the potential customers, advertisers use the above-mentioned vehicles. In order to maximize the effectiveness of advertising effort, media planning is the process of selecting time and space in various media for advertising. The best media plans provide the target audiences with an optimum level of coverage and opportunities to see the campaign. So, media allocation is to find the proper assignment of number of ads in each vehicle. This allocation problem can be developed as an optimization model, which can also be considered as the WTA problem of military operation research, that allocates media to target audiences.

This problem is an integer nonlinear programming problem which is independent of the duration of an advertising campaign also schedules advertisements during a day. This is an appropriate example of military operations research models that can be adapted to contemporary business world applications.

3.5 Various existing algorithms

Several exact and heuristic algorithms have been proposed to solve the Weapon-Target Assignment problem for several years. Some of them are described briefly below:

3.5.1 Maximum marginal return (MMR) algorithm

Maximum marginal return algorithms are algorithms that assign weapons sequentially with each weapon being assigned to the target which results in the maximum decrease (marginal return) in the objective function value. In other words, in maximum marginal return algorithms, a weapon is always assigned to the target with maximum improvement in the objective function value. Maximum marginal return algorithms are heuristic algorithms, they are easy to implement and efficient algorithms. Although these algorithms do not give the optimal or best solution it is known that these algorithms give near-optimal solutions.

Algorithm: MMR Algorithm

```
solution. Allocations← {}
solution. Value← MaxValue
allocated Weapon Count ←0
while allocated Weapon Count < =no Of Weapons do
    max Decrease ←Min Value
    k←1
   while k < unallocated Weapons. Count do
    i← 1
    while i < no Of Targets do
       decrease ←target Values [i] * kill Probabilities [i][k]
       if decrease > max Decrease then
          max Decrease← decrease
          allocated Target ← i
          allocated Target ← k
       end if
i← i + 1
    end while
       k← k + 1
   end while
   unallocated Weapons. Remove (allocated Weapon)
   solution. Allocations [k]←allocated Target
target Values [allocated Target] ← target Values [allocated Target]-max
   Decrease
   allocated Weapon Count← allocated Weapon Count + 1
end while
solution. Value← Calculate Solution Value (solution. allocations)
return solution
```

3.5.2 Genetic algorithms (GA)

A genetic algorithm with greedy eugenics that takes into account a probability of kill value for each weapon is suggested [14], and compared to MMR algorithm. Although MMR algorithm runs much faster than GA, GA tends to find better solutions than MMR algorithm. And, GA efficiency increases as the number of targets and weapons increases. Also in GA if a set of weapons can also hit a group of targets, meaning that grouping of weapons and targets is possible, this leads to faster and more optimal solutions [15]. Since the algorithm uses randomization it is a nondeterministic algorithm. The genetic algorithm is given as follows [16]:

Algorithm: Genetic Algorithm

```
start Time ←Now
end Time ←start Time + allowed Search Time
solution. Allocations← {}
solution.value← MaxValue
if no Of Targets > no Of Weapons then
no Of Individuals ←no Of Targets
else
no Of Individuals← no Of Weapons
```

```
end if
population← Generate Initial Population (no Of Individuals)
while end Time < Now do
individual No← 1
while individual No < =no Of Individuals do
    sol From Indv← population [individual No]
    sol Value From Indv← Calculate Solution Value (sol From Indv)
    if sol Value From Indv < solution. Value then
        solution← sol From Indv
    end if
    individual No← individual No + 1
    end while
    parents← Select Parents (population)
    population← CrossOver (parents)
    population← Mutate (population)
end while
return solution
```

3.5.3 Ant colony optimization algorithm

Ant-colony optimization takes inspiration from the foraging behavior of ant colonies. Initially all of the ants search for the food randomly. When an ant finds a food, it starts to deposit a chemical substance produced and released into the environment called pheromone on the ground while returning back to the colony. By depositing pheromone on the ground, they mark the path to the food that should be followed by other members of the colony. If an ant comes across a path with pheromone, it stops searching for the food randomly and starts to follow the path marked with pheromone. If it reaches the food, it starts to deposit pheromone on the path back to the colony also. This positive feedback strengthens the pheromone trail on the same path and causes all of the ants to follow a single path. On the other hand, if the path is not followed by other colony members, the pheromone evaporates in time and eventually, the path disappears [1, 16].

An Ant-Colony Optimization algorithm basically consists of 3 main steps. After the initialization of pheromone trails, while there is still time, at each iteration:

1. Ants create solutions.

2. Created solutions are improved through a local search. This process is also known as daemon actions and it is an optional process.

3. Pheromone update is applied to increase the pheromone values that are associated with good solutions and to decrease the pheromone values that are associated with bad solutions (pheromone evaporation).

The description of the algorithm is given below:

Algorithm: Ant Colony Optimization Algorithm.

```
start Time ←Now
end Time ←start Time + allowed Search Time
solution. Allocations← {}
solution. value← MaxValue
if no Of Targets > no Of Weapons then
```

```
  no Of Ants ←no Of Targets
else
  no Of Ants← no Of Weapons
end if
Calculate Heuristic Values ()
Calculate Pheromone Values ()
while end Time < Now do
  min Solution Value← Max Value
ant No← 1
while ant No < =no Of Ants do
    constructed Sol←Construct Solution ()
    if constructed sol.solution Value < min Solution. Value then
      best sol value← constructed Sol solution Value
      iteration Best Sol. Alloc← constructed Sol. allocations
      if constructed Sol. Solution Value < solution. Solution Value then
        solution← constructed sol
      end if
    end if
    Calculate Heuristic Values ()
      ant No← ant No + 1
    end while
    Update Pheromone Values (iteration Best Sol Alloc, best Sol Value)
    end while
return solution
```

3.6 WTA on real battle field

On modern battlefields, the task of battle managers is very important to make a proper assignment of weapons to targets to defend own-force assets or to offend the opponent targets. As an example, we now consider a target-based weapon-target assignment model for maximizing the total expected damage value of the targets which satisfies the Eqs. (3)–(5). Here considering five weapons are to be assigned to 20 targets [17, 18]. These targets have different probabilities of killing to platforms which are dependent on the target types. That is, the destroying probabilities of targets by different types of weapons obviously will be different. The probabilities define the effectiveness of the *ith* weapon to *jth* destroy the target. Here we get by the weapon-target pair that there are total 100 variables that are to be found out. The upper limits on weapon capacity and lower limits on weapons to be assigned are also given.

The characteristics of the five weapon types could be thought as follows:

1. Breda-SAFAT machine gun

2. Lewis gun

3. Spandau machine gun

4. Vickers machine gun

5. Blue Danube (nuclear weapon)

Each weapon-target pair survival probabilities are shown in **Figure 3**.

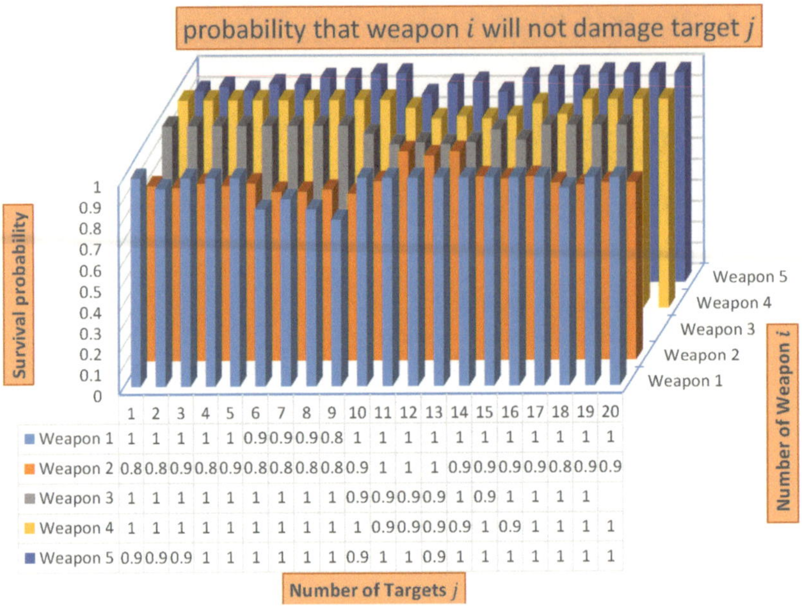

	1	2	3	4	5	6	7	8	9	10	11	12	13	14	15	16	17	18	19	20
Weapon 1	1	1	1	1	1	0.9	0.9	0.9	0.8	1	1	1	1	1	1	1	1	1	1	1
Weapon 2	0.8	0.8	0.9	0.8	0.9	0.8	0.8	0.8	0.8	0.9	1	1	1	0.9	0.9	0.9	0.9	0.8	0.9	0.9
Weapon 3	1	1	1	1	1	1	1	1	1	0.9	0.9	0.9	0.9	1	0.9	1	1	1	1	
Weapon 4	1	1	1	1	1	1	1	1	1	1	0.9	0.9	0.9	0.9	1	0.9	1	1	1	1
Weapon 5	0.9	0.9	0.9	1	1	1	1	1	1	0.9	1	1	0.9	1	1	1	1	1	1	1

Figure 3.
Survival probabilities of targets by weapons.

The number of available weapons and the military value of targets is shown in [19] **Figure 4**.

There are also some requirements for weapons to destroy particular targets. **Figure 5** shows the minimum number of weapons that must be assigned to some particular targets.

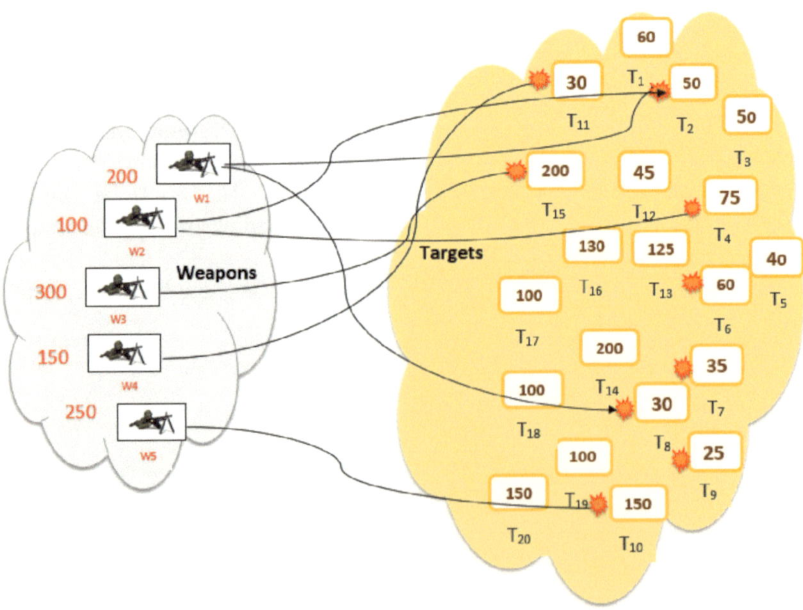

Figure 4.
Availability of weapons and target military values.

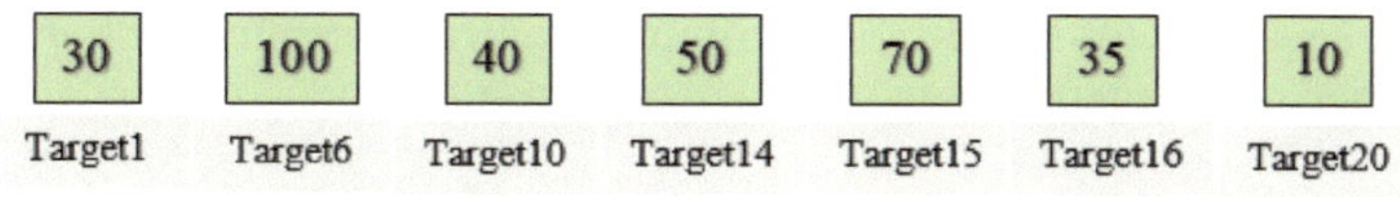

Figure 5.
Minimum requirements of weapons assigned to targets.

3.7 Formulation of battle field example

After having all the values of required parameters, we formulate the model corresponding to the given example for maximizing the total expected target damage value as follows:

$$
\begin{aligned}
z = {} & 60[1.00 - (1^{x_{11}} \cdot 0.84^{x_{21}} \cdot 0.96^{x_{31}} \cdot 1^{x_{41}} \cdot 0.92^{x_{51}})]+ \\
& 50[1.00 - (0.95^{x_{12}} \cdot 0.83^{x_{22}} \cdot 0.95^{x_{32}} \cdot 1^{x_{42}} \cdot 0.94^{x_{52}})]+ \\
& 50[1.00 - (1^{x_{13}} \cdot 0.85^{x_{23}} \cdot 0.96^{x_{33}} \cdot 1^{x_{43}} \cdot 0.92^{x_{53}})]+ \\
& 75[1.00 - (1^{x_{14}} \cdot 0.84^{x_{24}} \cdot 0.96^{x_{34}} \cdot 1^{x_{44}} \cdot 0.95^{x_{54}})]+ \\
& 40[1.00 - (1^{x_{15}} \cdot 0.85^{x_{25}} \cdot 0.96^{x_{35}} \cdot 1^{x_{45}} \cdot 0.95^{x_{55}})]+ \\
& 60[1.00 - (0.85^{x_{16}} \cdot 0.81^{x_{26}} \cdot 0.90^{x_{36}} \cdot 1^{x_{46}} \cdot 0.98^{x_{56}})]+ \\
& 35[1.00 - (0.90^{x_{17}} \cdot 0.81^{x_{27}} \cdot 0.92^{x_{37}} \cdot 1^{x_{47}} \cdot 0.98^{x_{57}})]+ \\
& 30[1.00 - (0.85^{x_{18}} \cdot 0.82^{x_{28}} \cdot 0.91^{x_{38}} \cdot 1^{x_{48}} \cdot 1^{x_{58}})]+ \\
& 25[1.00 - (0.80^{x_{19}} \cdot 0.80^{x_{29}} \cdot 0.92^{x_{39}} \cdot 1^{x_{49}} \cdot 1^{x_{59}})]+ \\
& 150[1.00 - (1^{x_{1,10}} \cdot 0.86^{x_{2,10}} \cdot 0.95^{x_{3,10}} \cdot 0.96^{x_{4,10}} \cdot 0.90^{x_{5,10}})]+ \\
& 30[1.00 - (1^{x_{1,11}} \cdot 1^{x_{2,11}} \cdot 0.99^{x_{3,11}} \cdot 0.91^{x_{4,11}} \cdot 0.95^{x_{5,11}})]+ \\
& 45[1.00 - (1^{x_{1,12}} \cdot 0.98^{x_{2,12}} \cdot 0.98^{x_{3,12}} \cdot 0.92^{x_{4,12}} \cdot 0.96^{x_{5,12}})]+ \\
& 125[1.00 - (1^{x_{1,13}} \cdot 1^{x_{2,13}} \cdot 0.99^{x_{3,13}} \cdot 0.91^{x_{4,13}} \cdot 0.91^{x_{5,13}})]+ \\
& 200[1.00 - (1^{x_{1,14}} \cdot 0.88^{x_{2,14}} \cdot 0.98^{x_{3,14}} \cdot 0.92^{x_{4,14}} \cdot 0.98^{x_{5,14}})]+ \\
& 200[1.00 - (1^{x_{1,15}} \cdot 0.87^{x_{2,15}} \cdot 0.97^{x_{3,15}} \cdot 0.98^{x_{4,15}} \cdot 0.99^{x_{5,15}})]+ \\
& 130[1.00 - (1^{x_{1,16}} \cdot 0.88^{x_{2,16}} \cdot 0.98^{x_{3,16}} \cdot 0.93^{x_{4,16}} \cdot 0.99^{x_{5,16}})]+ \\
& 100[1.00 - (1^{x_{1,17}} \cdot 0.85^{x_{2,18}} \cdot 0.95^{x_{3,17}} \cdot 1^{x_{4,17}} \cdot 1^{x_{5,17}})]+ \\
& 100[1.00 - (0.95^{x_{1,18}} \cdot 0.84^{x_{2,18}} \cdot 0.92^{x_{3,18}} \cdot 1^{x_{4,18}} \cdot 1^{x_{5,18}})]+ \\
& 100[1.00 - (1^{x_{1,19}} \cdot 0.85^{x_{2,19}} \cdot 0.93^{x_{3,19}} \cdot 1^{x_{4,19}} \cdot 1^{x_{5,19}})]+ \\
& 150[1.00 - (1^{x_{1,20}} \cdot 0.85^{x_{2,20}} \cdot 0.92^{x_{3,20}} \cdot 1^{x_{4,20}} \cdot 1^{x_{5,20}})]
\end{aligned}
\tag{7}
$$

subject to,
The linear constraints on the available number of weapons of the five types are,

$$
\begin{aligned}
& x_{11} + x_{12} + x_{13} + x_{14} + x_{15} + x_{16} + x_{17} + x_{18} + x_{19} + x_{1,10} + x_{1,11} \\
& + x_{1,12} + x_{1,13} + x_{1,14} + x_{1,15} + x_{1,16} + x_{1,17} + x_{1,18} + x_{1,19} + x_{1,20} \le 200 \\
& x_{21} + x_{22} + x_{23} + x_{24} + x_{25} + x_{26} + x_{27} + x_{28} + x_{29} + x_{2,10} + x_{2,11} \\
& + x_{2,12} + x_{2,13} + x_{2,14} + x_{2,15} + x_{2,16} + x_{2,17} + x_{2,18} + x_{2,19} + x_{2,20} \le 100 \\
& x_{31} + x_{32} + x_{33} + x_{34} + x_{35} + x_{36} + x_{37} + x_{38} + x_{39} + x_{3,10} + x_{3,11} \\
& + x_{3,12} + x_{3,13} + x_{3,14} + x_{3,15} + x_{3,16} + x_{3,17} + x_{3,18} + x_{3,19} + x_{3,20} \le 300 \\
& x_{41} + x_{42} + x_{43} + x_{44} + x_{45} + x_{46} + x_{47} + x_{48} + x_{49} + x_{4,10} + x_{4,11} \\
& + x_{4,12} + x_{4,13} + x_{4,14} + x_{4,15} + x_{4,16} + x_{4,17} + x_{4,18} + x_{4,19} + x_{4,20} \le 150 \\
& x_{51} + x_{52} + x_{53} + x_{54} + x_{55} + x_{56} + x_{57} + x_{58} + x_{59} + x_{5,10} + x_{5,11} \\
& + x_{5,12} + x_{5,13} + x_{5,14} + x_{5,15} + x_{5,16} + x_{5,17} + x_{5,18} + x_{5,19} + x_{5,20} \le 250
\end{aligned}
\tag{8}
$$

And the linear constraints on the minimum required assignment of weapons to the seven specified targets that must be engaged are:

$$
\begin{aligned}
&x_{11}+x_{21}+x_{31}+x_{41}+x_{51}\geq 30\\
&x_{16}+x_{26}+x_{36}+x_{46}+x_{56}\geq 100\\
&x_{1,10}+x_{2,10}+x_{3,10}+x_{4,10}+x_{5,10}\geq 40\\
&x_{1,14}+x_{2,14}+x_{3,14}+x_{4,14}+x_{5,14}\geq 50\\
&x_{1,15}+x_{2,15}+x_{3,15}+x_{4,15}+x_{5,15}\geq 70\\
&x_{1,16}+x_{2,16}+x_{3,16}+x_{4,16}+x_{5,16}\geq 35\\
&x_{1,20}+x_{2,20}+x_{3,20}+x_{4,20}+x_{5,20}\geq 30
\end{aligned}
\tag{9}
$$

3.8 Computational complexity of WTAP

The general WTA problem is the situation where a number of Wweapon systems have to engage a number of Ttargets. All weapon systems and all targets may have different characteristics. Also, different weapon systems may require a different amount of time to engage a target. When$T >> W$ an additional problem occurs. So as the scale of WTA problem grows, its computational requirement grows exponentially. So it is quite impossible to solve this type of large scale WTA problem directly. So, computational algorithms are the best approach to solve the large scale dynamic WTA problem [8].

3.9 Our solution approach for the WTA model

After formulating the problem, we have the Eqs. (7)–(9). We observe that we have total 100 variables with a nonlinear exponential objective function and 12 linear constraints, which is quite large. There does not exist any exact methods for the WTA problem even relatively small size problems. As there are so many computer based software tools to solve different types of mathematical problems, we propose a computer oriented algorithm to solve such large scale problems in a short time. Our proposed algorithm not only solve large scale WTA problems but also small problems in a single framework. We develop a computerized algorithm in which all types of target-based WTA problem can be solved in a reasonably fast time to help decision makers to make proper assignment on the battlefield.

3.10 The structure of our proposed algorithm for solving the WTAP

Since no real time exact solutions to WTAs are available, either for static or dynamic versions, alternative approximation methodologies must be considered, including heuristic techniques. We develop our computerized algorithm by using the Mathematical Programming Language AMPL.

Algorithm: Our developed Algorithm in AMPL.

Step I: *Initialize parameters N, M > 0 and set integers I, J, K.*

Step II: *Input number of weapons (M) and number of targets (N) and introduce non-negative integer variables*$x\{i,j\}$.

Step III: *Introduce the parameters* $w_i\geq 0, t_j\geq 0, u_j\geq 0, 0\leq p_{kill}^{ij}\leq 1$..

Step IV: *Define the non-convex objective function to maximize*

$$\sum_{j=1}^{J} u[j] * \left(1 - \prod_{i=1}^{I}\left(1 - p_{kill}^{ij}[i,j]\right) \wedge x[i,j]\right)$$

Step V: *Define a set of equations of constraints,*

$$\sum_{j=1}^{J} x[i,j] \le w[i]; \; i = 1, \ldots, M;$$

$$-\sum_{i=1}^{I} x[i,j] \le -t[j]; \; i = 1, \ldots, N;$$

Step VI: *Input data of the defined parameters in the 'dat' file.*

Step VII: *Then run this code in the 'run' file to calculate the objective function value that is to be maximized by using the solver option such as* ***MINOS, BARON, BONMIN, MINLP, and CONOPT.*** *By using the command* ***"EXPAND"*** *we can show the expansion of objective function and constraints in the console.*

Step VIII: *Display the solution value in the console.*

Using the new developed algorithm by AMPL, we can solve the WTAP for the large numbers of weapons and targets using the single model file with different data values according to the different scale problems.

3.11 Results of the WTAP using our developed method

As our developed method is based on computerized tools, so we first develop the general code in AMPL. Then update the data file for the Eqs. (7)–(9). And finally run the AMPL code, then we get our desired result as an output file (Appendix-A) in AMPL. Subsequent to adjusting the quantity of weapons to the closest whole numbers, the outcomes have appeared in.

3.12 Comparison between two results of the WTAP

For several years this type of weapon-target assignment has been performed at the Research analysis. Here we have taken the numerical problem presented in [18]. We have presented the result of the WTAP by using our developed method in **Table 2**. Bracken et al. [18] solved this problem and got a set of solution of the number of weapons assigned to targets shown in **Figure 6**.

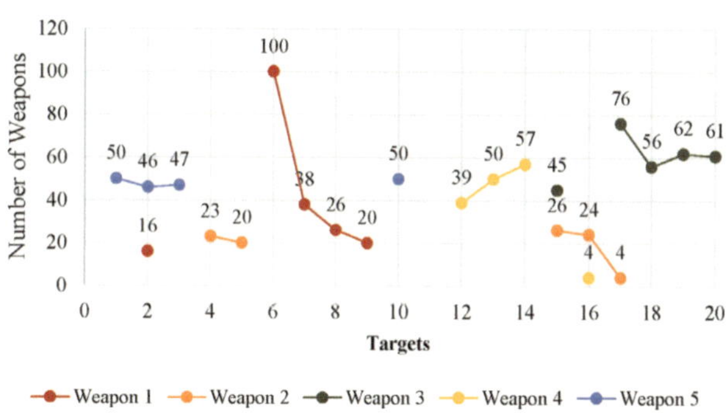

Figure 6.
Number of weapons assigned to targets.

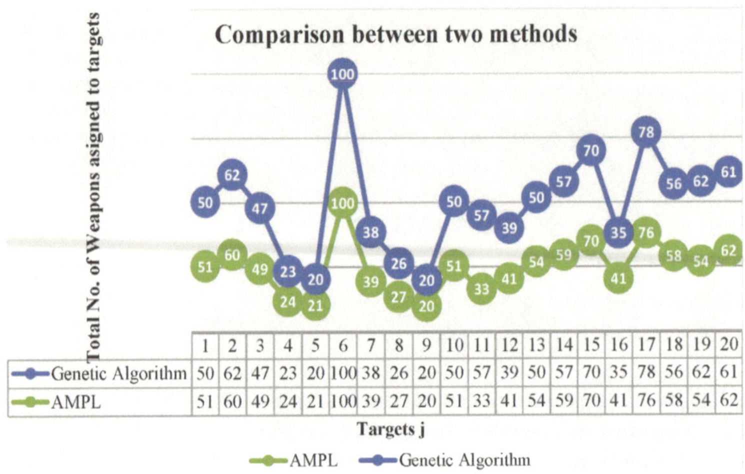

Figure 7.
Comparison of the results between the two methods.

Now to check the efficiency of our model we compare the two results of the problem graphically. Then for the both results we calculate the objective function that is to be maximized. Here the graphical representation of total number of weapons assigned to targets of the results are shown in **Figure 7**.

Objective Function (Existing Solution): Max $z = 1733.81$

Objective Function (Our Result): Max $z = 1735.57$

Comparing the above two results, we have the better result than the existing result. That is, we have the maximum objective function. This concludes that our proposed method gives the effective result. Our developed AMPL code studied in this Chapter improved the existing solution by 0.1%.

3.13 Numerical example of media allocation

Comparing media allocation with the WTA problem, we can consider the weapons as media vehicles to be advertised when the military targets as target audiences to be intended to reach. People exposed by media vehicles at different times of the day are given as target audiences. The weapon numbers x_{ij} are determined as the number of ads.

The number of ads refers to the number of times within a given period time an audience is exposed to a media schedule. The mathematical programming model is as follows under the assumption that the target audience is constant to be exposed by such media vehicles in given period time.

We formulate the media allocation as the weapon-target assignment model which satisfies the Eqs. (3)–(6).

Here $i = 1, 2, \ldots, W$ be the number of kinds of advertisements,

$j = 1, 2, \ldots, T$ be the number of segments,

w_i be the available number of advertisements of type i,

t_j be the minimum required number of ads for the target audience j s,

u_j be the relative segment weights.

x_{ij} be the number of advertisements of type i assigned to target j,

p_{kill}^{ij} be the probability of reaching the target audience j by a single ad type i.

So here the objective is to maximize the total probability of reaching the target audiences.

Suppose a company is planning to start an advertising campaign for a particular product. That company takes four target audiences as morning, afternoon, prime and night time of the day. Also, they take 15 vehicles such as somoy news, BTV, Channel I, NTV, ETV, ATN News, GTV, Radio Today, Radio Foorti, Facebook, Prothom Alo, Ittefaq, Billboard, Printings, and E-mail. That company knows the percentages of reaching the target audiences in different time partitions according to the mentioned media vehicles. The probabilities of reaching target audiences are shown in the following table. In **Table 2**, we can see that some vehicles have 0 probability to reach some targets. Prime time is the most important segment, as night time is the least important segment for the product. Moreover, the segment weights facilitate marketers to give relative importance with respect to product or service characteristics. The weights can be changed with respect to the features of the product.

Probability Matrix (p_{kill}^{ij}):

3.14 Formulation of media allocation problem

Our objective is to make a proper assignment of ads to targets for maximizing the effectiveness of advertising. The objective function along with total 19 constraints (15 supply constraints for media vehicles and 4 demand constraints for target audiences) are given below:

Maximize, z =

$$\begin{aligned}
&2[1.00-(0.79^{x_{11}}\cdot 0.65^{x_{21}}\cdot 0.81^{x_{31}}\cdot 1^{x_{41}}\cdot 0.87^{x_{51}}\cdot 0.76^{x_{61}}\cdot 0.91^{x_{71}}\cdot\\
&0.61^{x_{81}}\cdot 0.76^{x_{91}}\cdot 0.9^{x_{10,1}}\cdot 0.88^{x_{11,1}}\cdot 0.68^{x_{12,1}}\cdot 0.68^{x_{13,1}}\cdot 0.77^{x_{14,1}}\cdot 0.71^{x_{15,1}})]\\
&+3[1.00-(0.88^{x_{12}}\cdot 0.76^{x_{22}}\cdot 0.96^{x_{32}}\cdot 0.74^{x_{42}}\cdot 0.81^{x_{52}}\cdot 0.86^{x_{62}}\cdot 1^{x_{72}}\cdot\\
&0.83^{x_{82}}\cdot 0.69^{x_{92}}\cdot 0.77^{x_{10,2}}\cdot 0.89^{x_{11,2}}\cdot 0.77^{x_{12,2}}\cdot 0.90^{x_{13,2}}\cdot 0.88^{x_{14,2}}\cdot 0.93^{x_{15,2}})]\\
&+4[1.00-(0.88^{x_{13}}\cdot 0.88^{x_{23}}\cdot 1^{x_{33}}\cdot 0.81^{x_{43}}\cdot 0.75^{x_{53}}\cdot 0.78^{x_{63}}\cdot 0.82^{x_{73}}\cdot\\
&0.53^{x_{83}}\cdot 0.86^{x_{93}}\cdot 0.97^{x_{10,3}}\cdot 0.97^{x_{11,3}}\cdot 0.91^{x_{12,3}}\cdot 0.72^{x_{13,3}}\cdot 0.92^{x_{14,3}}\cdot 0.96^{x_{15,3}})]\\
&+1[1.00-(0.77^{x_{14}}\cdot 0.93^{x_{24}}\cdot 0.81^{x_{34}}\cdot 0.87^{x_{44}}\cdot 1^{x_{54}}\cdot 0.81^{x_{64}}\cdot 0.72^{x_{74}}\cdot\\
&1^{x_{84}}\cdot 0.57^{x_{94}}\cdot 0.65^{x_{10,4}}\cdot 0.91^{x_{11,4}}\cdot 0.79^{x_{12,4}}\cdot 0.98^{x_{13,4}}\cdot 0.97^{x_{14,4}}\cdot 0.68^{x_{15,4}})]
\end{aligned} \tag{10}$$

Subject to, the linear constraints on the available number ads of 15 media types are,

$$\begin{aligned}
&x_{11}+x_{12}+x_{13}+x_{14}\le 8\\
&x_{21}+x_{22}+x_{23}+x_{24}\le 7\\
&x_{31}+x_{32}+x_{33}+x_{34}\le 9\\
&x_{41}+x_{42}+x_{43}+x_{44}\le 5\\
&x_{51}+x_{52}+x_{53}+x_{54}\le 6\\
&x_{61}+x_{62}+x_{63}+x_{64}\le 8\\
&x_{71}+x_{72}+x_{73}+x_{74}\le 3\\
&x_{81}+x_{82}+x_{83}+x_{84}\le 10\\
&x_{91}+x_{92}+x_{93}+x_{94}\le 15\\
&x_{10,1}+x_{10,2}+x_{10,3}+x_{10,4}\le 12\\
&x_{11,1}+x_{11,2}+x_{11,3}+x_{11,4}\le 8\\
&x_{12,1}+x_{12,2}+x_{12,3}+x_{12,4}\le 4\\
&x_{13,1}+x_{13,2}+x_{13,3}+x_{13,4}\le 4\\
&x_{14,1}+x_{14,2}+x_{14,3}+x_{14,4}\le 4\\
&x_{15,1}+x_{15,2}+x_{15,3}+x_{15,4}\le 4
\end{aligned} \tag{11}$$

Media vehicles	Morning time (1)	Afternoon time (2)	Prime time (3)	Night time (4)	Ad capacities
Somoy News (1)	0.21	0.12	0.12	0.23	8
BTV (2)	0.35	0.24	0.12	0.07	7
Channel I (3)	0.19	0.04	0	0.19	9
NTV (4)	0	0.26	0.19	0.13	5
ETV (5)	0.13	0.19	0.25	0	6
ATN News (6)	0.24	0.14	0.22	0.09	8
GTV (7)	0.09	0	0.18	0.28	3
Radio Today (8)	0.39	0.17	0.47	0	10
Radio Foorti (9)	0.24	0.31	0.14	0.43	15
Facebook (10)	0.1	0.23	0.03	0.35	12
Prothom Alo (11)	0.12	0.11	0.03	0.09	8
Ittefaq (12)	0.32	0.23	0.09	0.21	4
Billboard (13)	0.32	0.1	0.28	0.02	4
Printings (14)	0.23	0.12	0.08	0.03	4
E-mail (15)	0.29	0.07	0.04	0.32	4
Number of ads required	16	18	25	10	
Segment weights	2	3	4	1	

Table 2.
The probability of reaching target audiences.

And, the linear constraints on the minimum required ads of media vehicles to the four specified target audiences that must be engaged are:

$$\begin{aligned}
&x_{11}+x_{21}+x_{31}+x_{41}+x_{51}+x_{61}+x_{71}+x_{81}+\\
&x_{91}+x_{10,1}+x_{11,1}+x_{12,1}+x_{13,1}+x_{14,1}+x_{15,1}\geq 16\\
&x_{12}+x_{22}+x_{32}+x_{42}+x_{52}+x_{62}+x_{72}+x_{82}+\\
&x_{92}+x_{10,2}+x_{11,2}+x_{12,2}+x_{13,2}+x_{14,2}+x_{15,2}\geq 18\\
&x_{13}+x_{23}+x_{33}+x_{43}+x_{53}+x_{63}+x_{73}+x_{83}+\\
&x_{93}+x_{10,3}+x_{11,3}+x_{12,3}+x_{13,3}+x_{14,3}+x_{15,3}\geq 25\\
&x_{14}+x_{24}+x_{34}+x_{44}+x_{54}+x_{64}+x_{74}+x_{84}+\\
&x_{94}+x_{10,4}+x_{11,4}+x_{12,4}+x_{13,4}+x_{14,4}+x_{15,4}\geq 10
\end{aligned} \tag{12}$$

3.15 Solution of the media allocation problem

We develop a near optimization model which allocate media vehicles to predetermined target segments. As this media allocation problem is formulated by using weapon-target assignment problem with 60 decision variables. By using our algorithm, we have solved the Media Allocation problem in a short time. In this case, we only change the data values in the 'dat' file, use the same mod.file and run. file. The result is given in **Figure 8**.

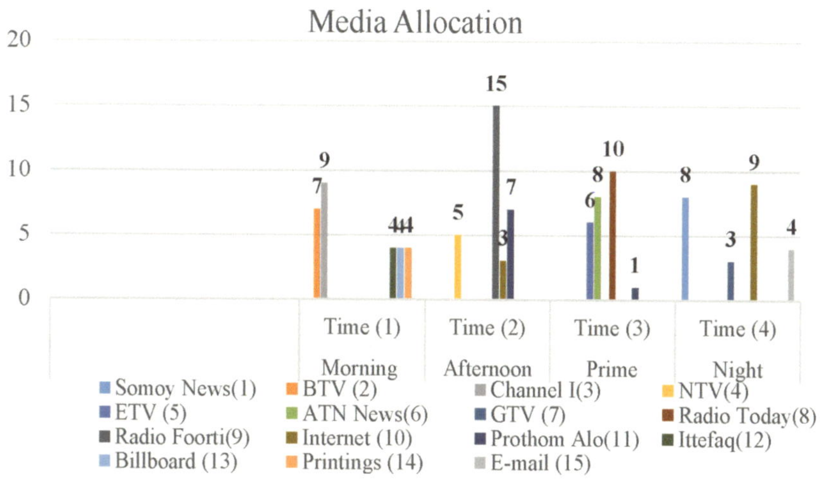

Figure 8.
Number of ads reaching to target audiences.

3.16 Comparison of media allocation result with other existing solution

This hypothetical example was given and solved by using MS Excel [20] and meta-heuristic genetic algorithm [21] previously. We have used our proposed algorithm to solve the media allocation problem. The solutions obtained by using genetic algorithm [21] and MS Excel [20] are shown in **Tables 3** and **4**, respectively [22].

To check the efficiency of our model, we need to calculate the objective function for all the existing solution that is to be maximized. So the graphical representation

Media vehicles	Morning time (1)	Afternoon time (2)	Prime time (3)	Night time (4)
Somoy News (1)	0	0	0	0
BTV (2)	7	0	0	0
Channel I (3)	6	1	0	2
NTV (4)	0	5	0	0
ETV (5)	0	0	6	0
ATN News (6)	2	0	5	0
GTV (7)	0	0	0	0
Radio Today (8)	2	0	8	0
Radio Foorti (9)	0	14	0	1
Facebook (10)	0	0	0	12
Prothom Alo (11)	2	2	2	2
Ittefaq (12)	1	1	1	1
Billboard (13)	1	1	1	1
Printings (14)	1	1	1	1
E-mail (15)	1	1	1	1
Total no of ads	23	26	25	21

Table 3.
Media allocation solution by genetic algorithm.

Media vehicles	Morning time (1)	Afternoon time (2)	Prime time (3)	Night time (4)
Somoy news (1)	1	0	0	0
BTV (2)	7	0	0	0
Channel I (3)	7	0	0	2
NTV (4)	0	5	0	0
ETV (5)	0	0	6	0
ATN news (6)	0	0	5	0
GTV (7)	0	0	0	0
Radio today (8)	2	0	8	0
Radio foorti (9)	0	15	0	0
Facebook (10)	0	0	0	12
Prothom Alo (11)	2	2	2	2
Ittefaq (12)	1	1	1	1
Billboard (13)	1	1	1	1
Printings (14)	1	1	1	1
E-mail (15)	1	1	1	1
Total no. of ads	23	26	25	20

Table 4.
Media allocation solution by MS excel solver.

of the existing solutions of Media Allocation and objective function value for the corresponding results is shown in **Figure 9**.

In **Figure 9** it is clear that, our model gives the best result compared to other two methods. By analyzing the values of the objective function, we can see that the Genetic algorithm improved the solution using MS Excel by 0.004%. Thus, the AMPL algorithm employed in this study improved the previous solution using Genetic Algorithm and MS Excel Solver 0.033% and 0.037% respectively.

In this effort, we have proposed the AMPL program code as a meta-heuristic tool for the solution of all type of dynamic weapon-target assignment problem. We have

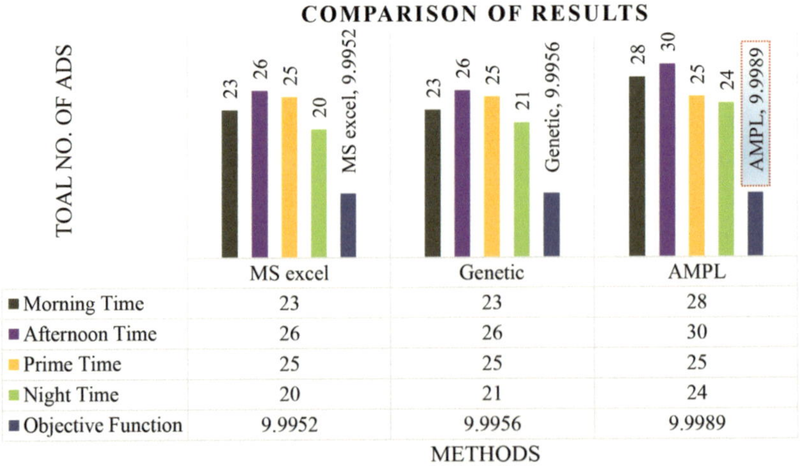

Figure 9.
Comparison the results between the three solving methods.

discussed two numerical examples and we have compared the results of the problems with previously solved results. We have observed that our proposed computational algorithm is easy to compute and gives a nearby optimal solution than other methods in a short time. We believe that AMPL program approach is a good and feasible alternative for the solution of this type class of problems. As further research, we may employ our developed AMPL program approach for the problem with many targets, many weapons or advertising tools as well.

4. Conclusions

This chapter is performed on two types of optimization such as the weapon's assignment problem.

In a warfare scenario, weapons allocation is very important. Since no exact algorithm is available to solve the WTAP, it is quite unavailable to estimate the quality of solutions produced by heuristic methods. The purpose of this chapter was to develop a new computerized algorithm to find a feasible solution in a reasonably fast time to help decision makers to make a proper assignment on the battlefield. We have developed a new computer-oriented algorithm by using AMPL to avoid the computational problems and solve this type of large scale problems. Our algorithm has been applied in two numerical examples of WTA problem and we have compared the complete outputs of the specified large scale problems with the outputs of the existing algorithms. We have concluded that our developed algorithm gives us the better result than others.

Finally, we conclude that the programming language AMPL is an effective technique to compute different types of optimization problems which will reduce the computational time for large scale problems. Overall, we have developed computer-oriented algorithms to solve the mentioned applications of optimization problems.

Author details

Mohammad Babul Hasan[1*] and Yaindrila Barua[2]

1 Department of Mathematics, University of Dhaka, Bangladesh

2 Daffodil International University, Bangladesh

*Address all correspondence to: mbabulhasan@yahoo.com

References

[1] Lee Z-J, Lee C-Y, Su S-F. An immunity-based ant colony optimization algorithm for solving weapon–target assignment problem. Applied Soft Computing. 2002;**2**(1): 39-47

[2] Manne AS. A target-assignment problem. Operations Research. 1958; **6**(3):346-351

[3] Day H. Allocating weapons to target complexes by means of nonlinear programming. Operations Research. 1966;**14**(6):992-1013

[4] Li Y, Kou Y, Li Z, Xu A, Chang Y. A modified Pareto ant colony optimization approach to solve bi-objective weapon-target assignment problem. International Journal of Aerospace Engineering. 2017;**2017**:1746124, 14 pages

[5] Tokgoz A, Bulkan S. Weapon target assignment with combinatorial optimization techniques. International Journal of Advanced Research in Artificial Intelligence. 2013;**2**(7):39-50

[6] Zhang Y, Rennong Y, Zuo J, Xiaoning J. Improved MOEA/D for dynamic weapon target assignment problem. Journal of Harbin Institute of Technology. 2015;**22**(6):121-128

[7] Hammond L. Application of a Dynamic Programming Algorithm for Weapon Target Assignment. Weapons and Combat Systems Division; 2016

[8] Li X, Zhou D, Pan Q, Tang Y, Huang J. Weapon-Target Assignment Problem by Multi-objective Evolutionary Algorithm Based on Decomposition, Complexity. 2018; **2018**:19

[9] Hosein PA, Athans M. Some analytical results for the dynamic weapon-target allocation problem. In: MIT Laboratory for Information and Decision Systems with Partial Support, Cambridge, MA, Tech. Rep. LIDS-p-1944. 1990

[10] Ma Y, Chou C. Weapon Target Assignment Decision Based on Markov Decision Process in Air Defense. In: Xiao T, Zhang L, Ma S editors. System Simulation and Simulation and Scientific Computing. ICSC. Berlin, Heidelberg: Springer; Vol. 327. 2012. pp. 353-360

[11] Boardman NT, Lunday BJ, Robbins MJ. Heterogeneous surfaceto-air missile defense battery location: A game theoretic approach. Journal of Heuristics. 2017;**23**(6):417-447

[12] Cai H, Liu J, Chen Y, Wang H. Survey of the research on dynamic weapon-target assignment problem. Journal of Systems Engineering and Electronics. 2006;**17**(3):559-565

[13] ACM, Karasakal O. Air defense missile-target allocation models for a naval task group. Computers & Operations Research. 2008;**35**(6): 1759-1770

[14] Lee ZJ, Su SF, Lee CY. Efficiently solving general weapon-target assignment problem by genetic algorithms with greedy eugencies. IEEE Transactions on Systems, Man, and Cybernetics. 2003;**33**(1):113-121

[15] Ahuja RK, Kumar A, Jha KC, Orlin JB. Exact and heuristic algorithms for the weapon-target assignment problem. Operations Research. 2007; **55**(6):1136-1146

[16] Lee Z-J, Su S-F, Lee C-Y. A genetic algorithm with domain knowledge for weapon-target assignment problems. Journal of the Chinese Institute of Engineers. 2002;**25**(3):287-295

[17] Mylander WC III. Applied mathematical programming. In: Proc. U.S. Army operations Res. Symp., Part 1. March; 1965

[18] Bracken J, McCormick G. Selected Application of Nonlinear Programming. New York: John Wiley & Sons; 1980

[19] Bogdanowicz ZR, Tolano A, Patel K, Coleman NP. Optimization of weapon–target pairings based on kill probabilities. IEEE Transactions on Cybernetics. 2013;**43**(6):1835-1844

[20] Çetin E, Esen ST. A weapon-target assignment approach to media allocation. Applied Mathematics and Computation. 2006;**175**(2):1266-1275

[21] Timur K, Cetin E. A genetic algorithm meta heuristic for the weapon-target based media allocation problem. Alphanumeric Journal. 2015;**3**(1)

[22] Yanxia W, Longjun Q, Zhi G, Lifeng M. Weapon target assignment problem satisfying expected damage probabilities based on ant colony algorithm. Journal of Systems Engineering and Electronics. 2008; **19**(5):939-944

Section 5

Emergence of Industry 4.0

Chapter 13

Automotive Industrial Supply Chain Performance Evaluation under Uncertain Constraints on Cloud Computing System

Suthep Butdee

Abstract

Performance evaluation is a critical and complex task as well as uncertain demands for automotive supply chain. Several methods are applied and adopted to deal with current situations and maintain competitiveness such as fuzzy logic, neuro fuzzy, agent (multi) based evaluation, etc. However, such systems are not rapid enough to respond customer requirements by real-time on mobile cloud computing system. There are many companies that operate under the first tier company as subcontractors on the same goal. Cloud computing system is capable to monitor real-time production processes for every subcontractor to assist the 1st tier to make decision and respond customer effectively. Daily monitoring data of all subcontractors in the supply chain are stored in the central database and finally the performance evaluation can be done. The implication is cost reduction of the whole supply chain and increase competitiveness as well as continuous process improvement for all.

Keywords: performance evaluation, automotive supply chain, uncertain constraints, neuro fuzzy system, CPOM system

1. Introduction

Supply chain management is emphasized on delivery parts on demands both regular, irregular, and uncertain circumstances. Inventory management is a critical aspect of management. There are four elements of inventory; raw materials, work in process (WIP), finished goods, and spare parts. The raw materials are components, subassemblies, or material that are purchased from outside the factory and used in the fabrication or assembly processes inside the factory. The WIP includes all finished parts or products that have been released to a production line. The finished good is final products or parts that are kept in the warehouse waiting to ship to customers by purchasing orders. Spare parts are standard components they bought from suppliers to maintain or repair production equipment. In automotive industry, a subcontractor is used as a major player in supply chain manufacturing cluster. The subcontractor is normally SME companies which makes contract for long period of time. Mostly, the subcontractor is sighed the contract according to the specific parts. However, the parts can be done by several contracts in order to reduce and distribute risky. The subcontractor can be classified into several levels namely 2nd tier, 3rd tier, and so on which relied on the role and performance evaluation. To become a subcontractor, a company is applied and audited by a SCM committee of the 1st tier company. The applicants are evaluated by company profile, factory visit, auditing 3 divisions; planning, quality assurance, and supply chain, based on the subcontracting criteria. There are 4 main criteria such as safety, planning, production control, and quality. The safety aims to check that the subcontractor works safely in the well-prepared environment including machine protection, operator protection, and part protection. Planning involves with production supporting process. They are production planning, raw material requirement planning, order receiving plan, compound and components receiving plan, sending raw material back to the 1st tier, delivery plan, packing standard plan, and inventory control. In terms of supply chain, It can be classified into 3 main group based on manufacturing types: sheet metal parts, plastic parts, and rubber parts. However, production technology is not concern in this chapter but it focuses on performance management and evaluation of delivery, back order clearance, and quality of finished parts. One of the most important aspects is that customer satisfaction on the customer service process that suppliers must provide any automotive parts needed whenever they want. In this paper, it is concerned on the case study and information based on the rubber part manufacturing industry. Production process control consists of compound storage control, curing and de-flashing, machine and mold preventive maintenance, and production control. Quality involves with inspection, finished goods and defect management as well as problem solving, quality assurance based on ISO (record, traceability, and change management).

The evaluation for renew-contract consists of two parts; monthly audit and yearly audit. The monthly audit is evaluated by the performance. They are mainly on delivery on-time, quality of finished goods, back order delivery. The result is achieved by grade level; A, B, C, and D. Grade A is excellent. B is good. C and D are needed to be improved. If the improvement is not successful and unsatisfaction, then the contract is resigned. The more details are in the later section. The yearly audit includes working system, ISO, improvement, control, and safety environment.

Supply chain performance evaluation for automotive supply chain is quite seriously emphasized on product quality and delivery on-time which are impact on human's live. The general problem is that the market is fluctuate and heavy competitiveness on the cost leadership. Supply chain and subcontractors are the key success methodology which are widely used all over the world. However, the problems are; 1. how to select subcontractors to produce parts effectively; 2. how to maximize cost when the capacities are suitably shared and distributes; 3. how to maximize cost for transportation of the whole subcontractor locations; 4. how to manage machine availability and uncertainty; 5. how to deal with uncertain performance and allocate production planning.

Modern computer information technology is currently applied to supply chain management such as computer integrated manufacturing (CIM), enterprise resources planning (ERP), internet of things (IOT), could manufacturing and so on. However, the SME subcontractors are not need to implement the fully high information technology or fully automation but it should invest or develop the parts of technology that can link to accompany the 1st tier company. One of the most critical part of the computer information technology used in the supply chain is could computing and monitoring as well as production execution in order to distribute and share information rapidly, real-time and on-time needed. Delivery is the most important for the automotive subcontractors especially at the 1st tier company. This chapter proposes automotive industrial supply chain performance evaluation under uncertain constraints on cloud computing system. Fuzzy logic based together with neural network or called neuro-fuzzy is developed to evaluate the performance of subcontractors and can be used for prediction the future events for rapid changes and provide a flexibility management on inventory and delivery to the end customer of OEM.

2. Concept and theory

This section presents the concept and theory of supply chain management, fuzzy logic, fuzzy AHP, Neural network and performance evaluation. Supply chain is major links to every parts of the business processes and communicate to their chains and the 1st tier in both vertical and horizontal organization. Fuzzy logic is a methodology used to deal with uncertainty including fuzzy AHP which helps to arrange the priority factors for multi-decision making to reduce complexity. Neural network is used to deal with qualitative data of performance evaluation.

2.1 Supply chain management

Supply chain plays significant role to OEMs in terms of increasing market share in a highly competition industrial cluster such as automotive industry by accessing to advanced manufacturing technology, reduced time to market, lower production costs, and more effectively used of assets. Modern supply chain management has seamless relationship in dealing with uncertain and fluctuate demands, risk management, stock management, allocate capacity, real-time tracking manufacturing progress with on-line and could system as well as using and selection outsource as engineering service providers. In addition, the customer relationship management is critical for the effective supply chain management. In the previous time, the average time required for company to process and delivery to market and customer from warehouse inventory taken many days or even unpredictable time because of uncertainty. It might be inventory out of stock, misplaced work order, misdirected shipment, total time to service customer escalated rapidly.

Presently, the world changing a lot influencing from digital age and disruption. Therefore, the information technology together with computerization are performed as a backbone of most of the business process according to 4G and 5G even the 6G is coming soon. Orders from customer are rapidly changed to e-commerce and on-line purchasing which is supported by modern logistics controlling by GPS and cloud monitoring. The reality of connectivity among collaborating business organization continues to drive a new order of relationship called supply chain management.

Supply chain is traditionally combined with logistic which consists of two major elements; inbound and out bound logistics. The process begins with second tier suppliers to deliver parts or products to the first-tier suppliers' through manufacturing processes. The inbound logistics are applied in this stage. Then, the finished goods are sent to distributors, agencies, and retails or even end users and customers. The structure of the supply chain is varied based on the companies. Therefore, the unique management is properly designed and created. However, the principle characteristics of supply chain are defined as ability of a firm to work with its suppliers to provide high quality material and components which are competitively priced. The closeness of relationship between vendor and customer in respects purchasing materials and inventory management reflects to the company's strategy and the role of supplier contributing to the long-term success of the firm.

Supply chain management is a well-known concept to applied in the modern business management all over the world particularly in the era of digital disruption and transformation. In the past, the supply chain management is applied to manage material flows among work stations inside the companies accompany with kanban, push-pull system, WIP and buffer inventory control in the JIT system including waste management of the lean manufacturing. Presently, the supply chain management is applied to a group of distributed manufacturing and outbound logistics in and out of the first company. It is recognized to be a critical tool to make survival and competitiveness even up to the competitive advantage of the whole chain. The key success factors are communication, delivery, quality, cost under uncertain circumstance. The digital transportation used is applied the progress of manufacturing processes by on-line and real- time tracking using cloud computing system. The evaluation can be measured by the key success factors in order to achieve the customer satisfaction. On the other hand, the supply chain cluster has to implement lean production in order to dealing with competitiveness problem. Company often had an antagonistic relationship with their suppliers inside the supply chain cluster. Every item that was purchased had several vendors who are played off against each other in order to obtain the lowest possible prices, which was the first criteria for being awarded a contract. Subcontractors recognizing that the relationship could very likely be terminated with the next contract, invested minimal time and money to address the specific needs of individual customers. The purchasing function within the manufacturing company often reported to the supply chain manager in order to make decision on purchasing raw material and components at the lowest possible cost.

2.2 Performance evaluation

Performance of production is the final goal of manufacturing which needs to measure for evaluation at the end of processes or periods of time. It can consider on quantity, quality, cost, value, productivity, resource used and so on. Effective performance evaluation for competitive advantage particularly in supply chain logistics includes monitoring, controlling, and directing operations. Monitoring is accomplished by the establishment of appropriate metrics to track system performance for reporting to management such as on-time delivery, production rate, production quality on planned, finished goods, defects, transportation tracking, warehousing and so on [1]. Controlling is accomplished by having appropriate standards of the whole life cycle of operation performance related to the customer requirements or international standard (ISO). Controlling is also used to ensure that planning time is managed effectively on the productive operations. Directing is related to shop floor control on the machine level and operators to run the production properly, correctly, and perfectly in order to achieve high level of productivity.

Performance metrics is typically involved with several criteria; cost, customer services, quality, productivity and asset. Costs are cost per unit, total cost, percentage cost of sales, administrative coat, direct labor cost, inventory carrying cost, cost of return goods, cost of defect, cost of damage, cost of service failures, cost of back order, cost of logistics, cost of materials and so on. Criteria of customer services are on-time delivery, back orders, cycle time, complete orders, delivery consistency, fill rate stockouts, response accuracy, customer complaints, reliability, overall satisfaction. Quality is the most criteria consisting of product quality, order entry accuracy, invoicing accuracy, information accuracy, number of customer complaints, number of customer returns, picking and shipping accuracy. Productivity criteria are decrease cost rate, number of increase production quantity rate, productivity index, order per sales representation, units per labor values, units shipped per employee,

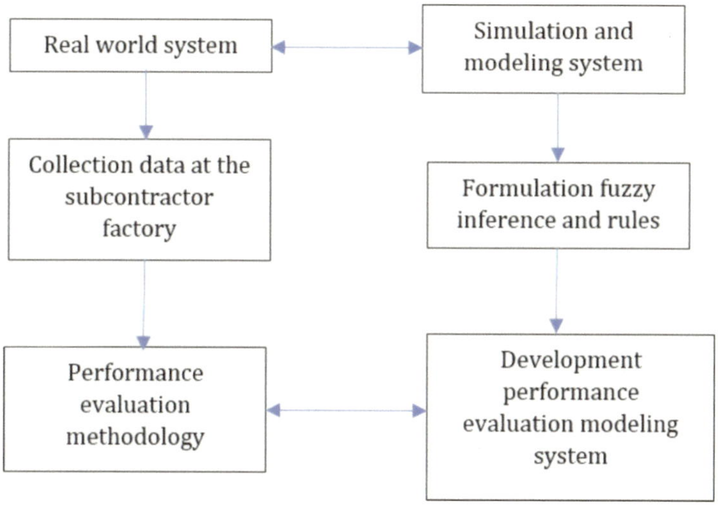

Figure 1.
Simulation and modeling framework for performance evaluation.

decrease of standard time, reduction of equipment downtime, warehouse and transportation labor productivity. Asset is measured by inventory turns, inventory levels, number of days of supply, obsolete inventory, return on net assets, return on investment, economical value added (EVA) and so on.

Performance evaluation for supply chain is different from a common firm particularly for subcontractors. There are multi-criteria evaluation in several hierarchy and different weights depending of the importance priorities. The main criteria are delivery on the purchasing orders and on-time and on quality. If any uncertainty is occurred on the delivery, back order must be released on the certain time. The quality involves with defects and errors occurring on the delivery. It is normally measured by six sigma (ppm). Presently, modeling and simulation are used as a manufacturing system tool for production flow and performance evaluation that most of the mathematical models are used in operations management and industrial engineering.

Figure 1 shows the simulation and modeling which applies fuzzy logic and Neuro fuzzy to evaluate the performance of subcontractors in automotive supply chain. It is started on the real-world system. The objectives, constraints, and alternatives are identified. Customer requirements are taken into account for designing and creating measurement parameters and controls for modeling using mathematic model or intelligent system such as fuzzy logic, neural networks and so on. Data at the current and activated factory of the subcontractors while doing real business is collected. Then the performance evaluation methodology is designed and implemented. On the other hand, the simulation and modeling system are developed. Fuzzy logic and neuro fuzzy system are formulated with membership functions, fuzzification and defuzzification. Finally, the performance evaluation modeling system are created and implemented.

3. Related works

This section explains the previous works on the methodology dealing with performance evaluation in the uncertainty circumstances because of changing world caused by modern communication and management of digital technology. Fuzzy logic is firstly explained following with neuro-fuzzy system. Then previous works on performance evaluation particularly on supply chain management is reviewed and comments. Finally, the modern management on cloud monitoring is stated and discussed to be the way of applying to the monitor performance operation at shop floor level.

3.1 Fuzzy logic

Fuzzy logic is very useful tool to support decision making under uncertainty and complexity when information is not sufficient and imprecision. Traditionally, probability theory is used to deal with uncertainty. However, the uncertainty can be different forms. The probability can deal with the expectation of the future event based on something known before. By contrast, fuzzy logic handles with a prediction about the event which represents fuzziness expression in terms of linguistic [2]. Zadeh [3] first presented a Fuzzy Set theory. It is a tool that helps in decision making under the uncertainty of data. The fuzzy logic has flexibility and can make easy decision making by applied in the representation of human reasoning and linguistic terms with the following definitions:

3.1.1 Definition 1. Fuzzy set

A fuzzy set is a member function that attributes the elements of the domain in the interval [0,1] as the following Eq. (1). The interval is a degree of membership when the given value is close to 1, there will be a higher level of membership. If the membership is zero, it means that there is no membership.

$$U_a(X) : U \rightarrow [0,1] \tag{1}$$

In which U_a (x): U → [0,1] is called pertinence function and U_a (x) is the degree of pertinence of x.

3.1.2 Definition 2. Linguistic variable

A Fuzzy set can be used to describe the value of a variable. A linguistic variable is a linguistic expression which is used to determine the value of what is described in terms of qualitative scales such as very low, low, medium, high and very high.

3.1.3 Definition 3. Fuzzification

The fuzzification is used to convert the input into fuzzy variables or fuzzy sets or language variables.

3.1.4 Definition 4. Membership function

The membership function is the process of determining the membership level of a variable, which is important for the process of thinking and solving problems. The membership functions are not symmetric or symmetrical in all respects. The membership function used in this paper is triangular. The triangular membership function has a triangle shape, which depends on the 3 variable values, a, b and c as shown in **Figure 2**. It is a commonly used in many researches.

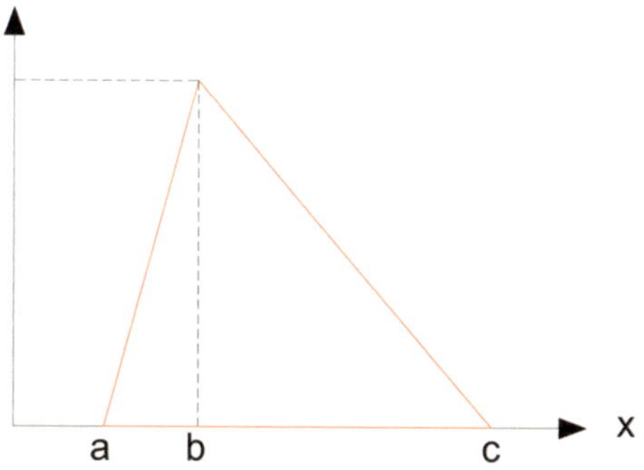

Figure 2.
Triangular membership function.

3.1.5 Definition 5. Defuzzification

A defuzzification operation is a process to transform fuzzy set into a crisp output. The center of gravity (COG) is the most commonly technique calculating the center of the area of combined membership function. In this paper, fuzzy logic sugeno type was used to approximate output as the following Eq. (2):

$$y = \frac{\sum |y = y'| \times y'}{\sum |y = y'|} \tag{2}$$

Previous works have been done using fuzzy logic for operation performance evaluation particularly in the area of supply chain management. Fuzzy logic for logistics management using SCOR model for evaluating performance in supply chain is widely studied [4–6]. The model developed can evaluate overall performance by real-time network. The model can predict the performance based on causal relationship of the SCOR metrics using fuzzy rules to build the prediction model. The fuzzy logic model can be combined with Linear Programming (LP) which aimed to evaluate the performance and capability in order to distribute purchasing orders in the cluster environment. The study presented the cluster body manufacturing in Thailand [7]. The cluster consists of 40 SME companies which located in the same areas. The have limitation of manufacturing resources. Therefore, they have to combine capacity in order to serve the big lots of purchasing orders. It is normally order with big lots of busses particularly the order from local government. For example, the 2016 purchasing orders were around 500 city busses with limited delivery time. The cluster of bus body manufacturing needs to join and share capacity, resources and even profit. The model develop was tested and help lots of benefits to them. The KPI indicators of the efficiency performance evaluation which is designed with 5 criteria; capacity, quality, reliability, flexibility and source. The fuzzy logic is extended to integrated with other methods such as Quality Function Deployment (QFD) to use for quality design of large complex products and structures. Although the fuzzy logic is effective but it has got a weakness in term of prediction by learning. In this case, the Artificial Neural Network (ANN) can be done and substitute the fuzzy logic.

3.2 Artificial neural network and neuro fuzzy system

ANN mimics human's brain working using algorithms and graph theory. It represents biological nervous systems such as brain processing information deriving from imprecise and complicated data. Training ANN can be used to provide projections given new situations [8]. Kocamaz et al. [9] proposed ANN which was used to control chaotic supply chain based on a nonlinear dynamic system. It is sensitive dependence on initial conditions and involves with infinite number of different periodic responses. Zayegh and Bassam [10] presented neural network principals and applications. ANN can be used to implement different stages of processing systems based on leaning algorithms by controlling their weights and biases. The paper presented the ANN application for digital signal processing performed by the concept of a multilayer perceptron which feeding forward as networking system by a set of neurons together with weights. It consists of an input layer, multi hidden layers and output layer. Back propagation is an algorithm working together with the multilayer perceptron as shown in the **Figure 3**.

Feng et al. [11] proposed the application of ANN to solve the problem of job shop scheduling using MLP networks. Job shop scheduling involves with several tasks varying from customer orders. A job shop consists of many jobs which are assigned to perform on many machines. ANN is advantage on learning by training approach and mostly deal with qualitative information. However, there are various types of application in manufacturing need to deal with the combination of quantitative and qualitative information. Previous articles stated that neuro-fuzzy can deal with the combination information effectively. There are many applications of ANN to production manufacturing such as Lee and Shaw [12] applied a neural network to deal with a real-time floe shop sequencing. Che [13] proposed the ANN for estimation cost of plastic injection molding. Efendigil and Onut [14] proposed the neuro-fuzzy based methodology to make multi-decision on multi-stage supply chains which integrates from customers to suppliers through warehouses, retailers, and factory. The ANFIS is used for making decision as artificial neural fuzzy inference system on the customer demands. It consists of several procedures; fuzzification, rule antecedent, normalization, rule consequent, and rule interface. The output carries out with new demand data. Chupin et al. [15] proposed neuro-fuzzy model in supply chain management for object state assessing in the conditions of uncertainty-based supply chain strategy. Didehkhani et al. [16] presented assessing flexibility in supply chain using adaptive neuro fuzzy inference system in order to consider the factors of competitiveness of the world class manufacturing. The criteria include speed, variety, flexibility and integration in production main line. Sremac et al. [17] proposed neuro-fuzzy inference systems approach to decision support system for economic order quantity in supply chain management based on a dynamic situation of information flow, products and funds among different participants. SCM is a complex process and mostly involving uncertainty.

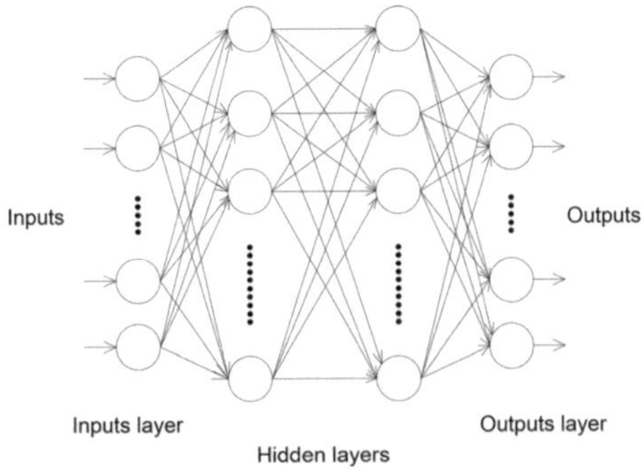

Figure 3.
A multilayer perceptron of neural network [11].

Thipparat [18] proposed application of adaptive neuro fuzzy inference system in supply chain management evaluation in the case study of construction project. The project consists of design, contract, liabilities, weather, soil conditions, environment, and so on which are uncertain. The construction project often deals with many parties of stakeholders in supply chain. The criteria are taken into account such as cost, asset, flexibility, reliability, responsiveness.

3.3 Sub-contractor evaluation

Evaluation is needed to all subcontractors every month and every year in order to check capability, performance and availability to continuous the business. The major criteria are delivery finished goods and all item according to the purchasing order. The 1st tier company plays responsibility to the OEM which is to delivery any part their needs on demands uncertainly. The facing troubles are that demands are fluctuated based on the world economic situations. Therefore, the real-time monitoring is needed in order to dealing with rapid change and effective control stock of inventory. Mahmood [19], Kaganski [20], Amrina & Yusof [21] proposed the concept of production evaluation in SME network using virtual enterprise. Key performance indicators (KPIs) is created which affect operation such as efficiency, utilization, and productivity. Hon [22] proposed the KPIs of production performance consisting of 5 criteria; quality, cost, delivery, and flexibility. Performance evaluation with KPIs is a good tool for controlling and enhancing the company to improve productivity and competitiveness but the methodology cannot use the same criteria particularly for evaluating a supply chain performance. Behrouzi et al. [23] presented the performance evaluation for supply chain management in the automotive industry. It consists of 20 KPIs which applied lean concepts in the context. The studied collected data from the 133 supply chain companies. The criteria contain attributes; waste elimination, continuous improvement, just in time, and flexibility. Roda & Macchi [24] proposed an evaluation model for production system based on the need of factory level performance metric tracking system which adopted OEE criteria to create KPIs. Joppena et al. [25] introduced a KPIs for production performance evaluation consisting of 5 criteria; quality rate, manufacturing defect rate, rework rate, rejection rate and OEE. Hyytia and Rgihter [26] proposed a performance evaluation using simulation modeling for dispatching systems of the routing jobs to the work stations using parallel computing systems. The system can deal with dynamic dispatching policy accompanied with monte Carlo methodology. In supply chain, the problem sometimes needs to deal with capacity sharing in order to increase potential a large order quantity which is benefited to SME industrial cluster. Butdee and Nitnara [8] proposed a fuzzy logic combined with linear programming (LP) modeling for the cluster capacity and performance evaluation to distribute purchased order suitably for each supply chain cluster of bus body manufacturing in Thailand. The modeling can deal with uncertainty situations when demands are fluctuated. The criteria include capacity, quality, reliability, sources, and flexibility. Jagan et al. [27] reviewed a supply chain performance of the whole production system using the concept of Balance Score Card combined with SCOR model as well as AHP model. Hudson et al. [28] proposed the theory and practice of performance measurement systems for SMEs using multi-criteria such as quality, flexibility, time, finance, customer satisfaction, and human resources. Svalina et al. [29] proposed a neuro fuzzy system for evaluation surface roughness with minimize machining time and maximize material removal rate, recommendation optimal cutting parameters under alternated possibility controlling of the machining processes. The system can deal with complexity, imprecision and uncertainty environment. The developed AHP and Fuzzy AHP modeling to deal with constraints and dynamic situations as well as uncertainty risk management.

Performance evaluation is currently involved with energy consumption which is enable to save production cost. Energy evaluation is concerned with the energy used for machines, oven, welding machines and so on. The energy consumption management is taken into account together with manufacturing efficiency. Energy is one of the major costs of automotive supply chain particularly the rubber part manufacturing. Most of the compression machines, injection machines are used the heaters to heat rubber materials and continuously heated whole day and whole weeks. It is sometimes worked around 20 days or more per month. Therefore, the energy is critical for the company in supply chine because cost is the major criterion of the competition. Energy control is dynamic and uncertain in aspects of obtaining the OEE of s firm. Scientific methods are applied to deal with the energy control. AHP and Fuzzy AHP including Fuzzy FMEA or even the combination of them. The criteria concern with the cause of uncertainty and risk such as severity, occurrence, and detection. The major goals of production control are cost control, improving quality, and on-time delivery.

3.4 Cloud management system

Cloud manufacturing system is emerged which is developed from the advanced manufacturing systems using the information and computer technologies such as cloud computing, internet of things (IOT), virtualization and service-oriented technologies, virtualization, cyber physical system. It transforms manufacturing resources and manufacturing capabilities into manufacturing services, which can be managed and operated in an intelligent and unified way to enable sharing facilities. The cloud manufacturing system can provide efficient and reliable, high quality, cost saving control. The system can enhance in the decision-making process while the material flow is tracked following the production process particularly at the final operation of the supply chain which needs to control on-time delivery. Problem solving can be rapidly done even immediately to increase productivity. The collaborative manufacturing management can be done along the whole life cycle of production. Few papers studied and presented could computing, cloud-based infrastructure. Lin Zhang et al. [30] expressed the cloud manufacturing which is the combination of cloud computing, IOT, serviced-oriented technologies and high-performance computing in order to solve the bottlenecks in the informatization development and manufacturing applications. It consists of three core components such as resources, cloud service and manufacturing on cloud. The evolutionary of modern manufacturing started from computer integrated manufacturing system (CIM). Then, the agile manufacturing is established to deal with time to market and supply chain management. It can support the responsiveness quickly while cost and quality can be consistently controlled. Afterward, concurrent engineering and collaborative design are developed to save time to market which was the life cycle of product taken into account. Peng Wang [31] explained the benefits and limitations of could computing for cloud manufacturing. Cloud computing is recognized at the moment that is an aggregate computing resources and a service of things. Presently, services on demand is dramatically increased when the world is in the digital disruption and transformation. Cloud computing is enabled to transition of computation in various forms such as platform, hardware, and software. The benefits of the cloud computing are to obtain lower start-up and operating costs, to ease of access and scalability, to reduce risk on resource provision.

The characteristics of cloud management system are to envision for any feature of IOT such as self-service, visualization of system image, workload optimization, interfaced tool management, service catalogs, network and storage configurations, and service governors, high performance management [32]. Self-service is the portal that interfaces to log and manage on infrastructure easily for configurations and execution deployments via a templates and customization. Visualization system image is enabled you to access different kind of system images that can be chosen and deployed via the self-service feature. Workload optimization can link to work anywhere seamlessly. It can work automatically or support decision making. It can manage resources better under policy and management rules. For example, the system server is used up to 60% or 75% with 2 GB of the ram. Interface to other management tools is to be a centralization of the data for monitoring overall information and work flow and deployment automation both existing technology and new equipment. This feature can assist the potential of execution system. Service catalogs is allowed users to choose service templates or configurations to apply any service in the catalogs via a list of promotion packages. In addition, users can create and developed their own template in different service categories. Network and storage resource configuration permit users to define several types of storage and networks even set up new storage configuration and sharing bloc level. Service governors are a smart tool which help users understand IT system inside company. The cloud has to support high performance management for gathering and collecting policies from the services based on infrastructure pattern use. It can configure automatically and deploy services using the right amount of resources and give high performance.

There are a lot of application using cloud system in different domains; automotive industry for process improvements and cost reduction, pharmaceutical production data for down time analysis, medical devices monitoring, education and training, production data for real time monitoring, food and beverage for OEE analysis to increase productivity.

4. Research system methodology

4.1 Subcontractor operation

Subcontractors are operated under the supply chain management which are closely linked to the 1st tier company. The most important aspect is that the subcontractors must understand the role of collaborative working. The main responsibility of the subcontractors is to receive purchasing order and delivery plan. The **Figure 4** shows the cycle of operations. It starts from receiving parts and delivery order. The raw material requirement planning is done by subcontractors. The production plan is created immediately after finished good stock is checked. Raw material is critical aspect when synthetic rubber is taken into account because its life is limited. It needs to be well-organized and can deal with uncertain environments. The raw material is ordered from the 1st tier company which all volume is combined from several subcontractors when the same material is used. This is the first advantage of the subcontractor supply chain which can share critical scale of resource by obtaining cost reduction. The production can start suddenly when the raw material is arrived. However, the material needs to test beforehand according to the standard specification and operation standard procedure given by the engineering department. The production consists of three operations before delivery to the customer of the 1st tier according to the delivery plan. The production also must be well-prepared for preventive maintenance both machines and molds. The molds are design accompanied with new part model before they are distributed to the subcontractors. This point is the 2nd advantage of the subcontractor supply chain

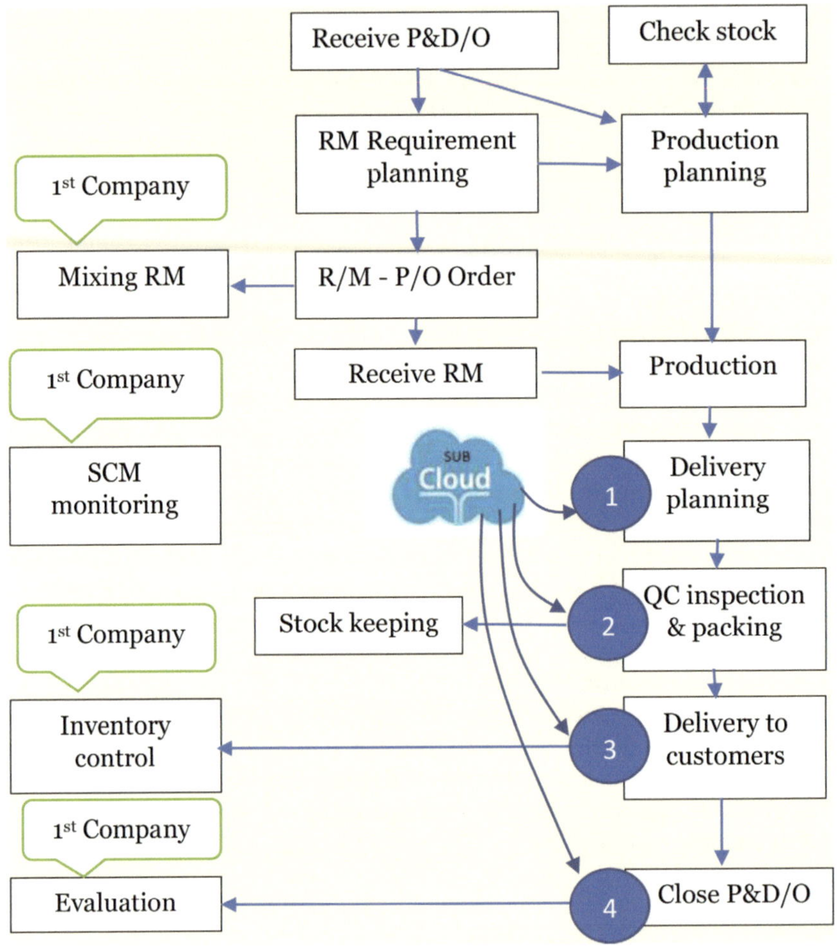

Figure 4.
The cycle of subcontractor in automotive supply chain management.

method. It is not only cost reduction but also the quality improvement. The delivery plan is monitored by the SCM department of the 1st tier following with the QC inspection and packing progress in order to predict the efficiency of the supply chain management. All of the parts required on the exact due date are delivered to the 1st tier inventory which is controlled by RFID and barcode inside the centralized data based and linked to the OEM. The final step is to close the P&D order and evaluated the performance. The concept of CPOM, detailed in 4.3, system is applied to the 4 last steps of the SCM. First is the delivery plan. Then, the QC& packing work station and stock control. The 3rd operation is delivery parts to customers which is linked to the 1st tier inventory control. The last operation is to close the P&D order which is linked to evaluation system.

Figure 5 shows production flow for supply chain collaborative of the 1st tier and the subcontractors. Based on the functional department. There are 5 departments which are defined in the same supply chain; engineering, planning supply chain, subcontractor and QC. The P&D order starts from the planning department and sends to the supply chain department and passes to the subcontractor. The engineering department prepares molds and operation standard procedure which is obtained from the new model testing at the real environment. The mold is sent to the subcontractor and tested for completion quality. Then, the subcontractor begins the preproduction and first lot experiment. The result is recorded and sent to the

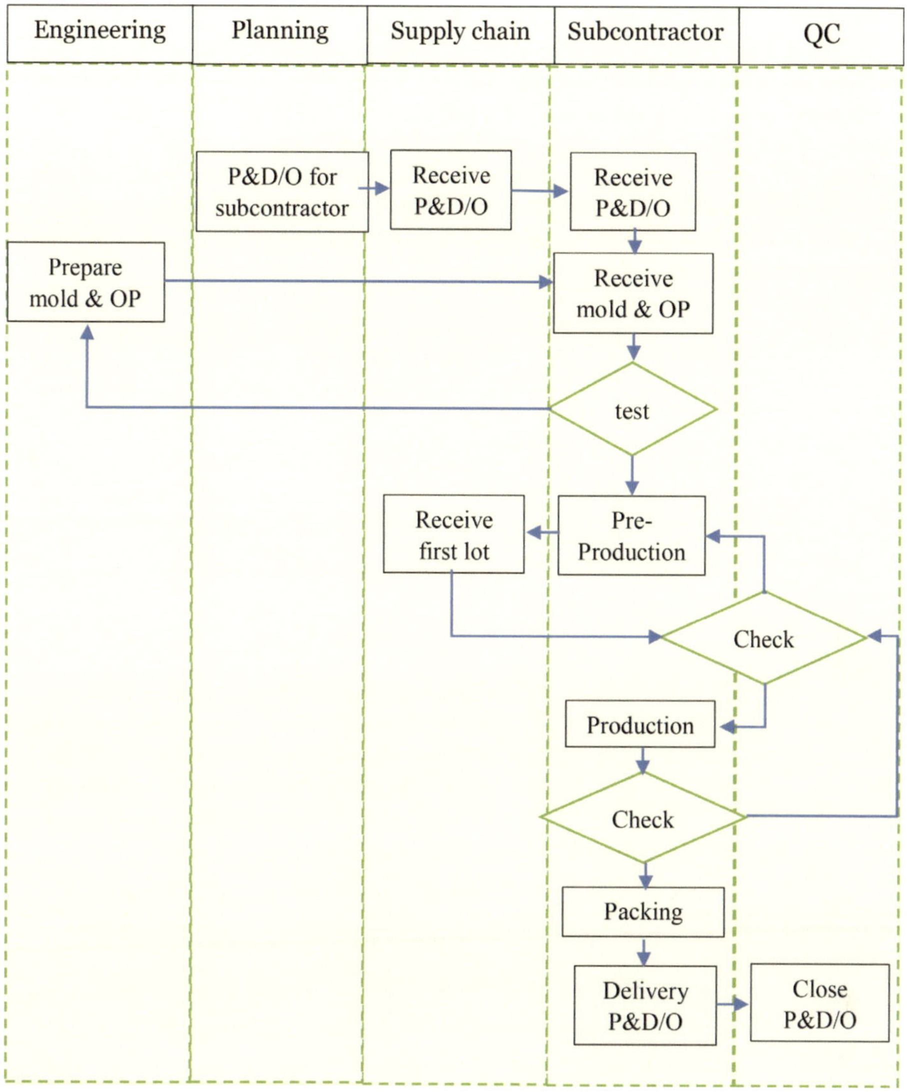

Figure 5.
Production flow for supply chain collaborative of the 1st tier and subcontractor.

QC department at the 1st tier to be approved. The production is continuing if the first lot is confirmed to be acceptable parts. The finished goods are checked by QC and packed before sending to the inventory, and close the P&D order.

4.2 Neuro fuzzy subcontractor evaluation system

This section presents the methodology of Neuro Fuzzy subcontractor evaluation before sending to the cloud production online monitoring (CPOM). This method is successfully validated and workable for automotive rubber subcontractor. The criteria include capacity, availability, quality, delivery, productivity, back order control. This study concerns only the monthly audit and yearly evaluation. As previous mentioned, the monthly is important because it affects the daily production performance. It is weight 70% of the full mark of the total evaluation procedure. Therefore, the yearly evaluation weight 30%. The subcontractors are evaluated with the same criteria and get results in 4 levels of satisfaction; A, B, C, and D. The grade A and grade B are passed with satisfaction whereas the grade C and the grade D are fail but they can improve the performance under the time and quality condition. There are 4 criteria are measured; 1. delivery on-time; 2. on quantity of the P&M order; 3. quality of the parts without any defect; 4. Back order clearance. The calculation performance is divided into three categories; delivery (D), back order clearance (B), and quality (Q). The function is shown in the Eq. 1, Eq. 2, and Eq. 3. Each criterion is applied and evaluated by every part. The performance evaluation level is given by score.

$$P(D) = \sum_{i=1}^{i=n} Di \sum_{j=1}^{j=n} Pj.(it1 + it2 + it3 + \cdots itn) \tag{3}$$

$$P(B) = \sum_{i=1}^{i=n} Bi \sum_{j=1}^{j=n} Pj.(it1 + it2 + it3 + \cdots itn) \tag{4}$$

$$P(Q) = \sum_{i=1}^{i=n} Qi \sum_{j=1}^{j=n} Pj.(it1 + it2 + it3 + \cdots itn) \tag{5}$$

When.

P(D) = delivery performance (%), P(B) = back order performance (%), P(Q) = quality performance (%), it = items to be produced (item consists of many pieces), Di = number of items to be delivered, Bi = number of back order to be delivered, Qi = quantity to be produced, Pj = number of pieces to be produced

$$P(D) = \begin{cases} [x|x \text{ (delivery performance)} \geq 99\%; \text{to get 15 marks}] \\ [x|x \text{ (delivery performance)} = 90\text{-}98\%; \text{to get 12 marks}] \\ [x|x \text{ (delivery performance)} = 80\text{-}89\%; \text{to get 9 marks}] \\ [x|x \text{ (delivery performance)} = 70\text{-}79\%; \text{to get 6 marks}] \\ [x|x \text{ (delivery performance)} < 70\%; \text{to get 0marks}] \end{cases}$$

The function set of P(D) shows the performance calculation of delivery which infers that the subcontractors have capability to handle the routinely received orders. The score is divided into 5 levels. The level 1 is excellent which can deliver more than 99% of the total order and pieces and get 15 marks. The level 2 is very good which can deliver between 90 and 98% of the total order and pieces and get 12 marks. The level 3 is good which can deliver between 80 and 89% of the total order and pieces and get 9 marks. The level 4 is moderate which can deliver between 70 and 79% of the total order and pieces and get 6 marks. The level 5 is fail which can deliver less than 70% of the total order and pieces and get 0 mark. The function set of the back-order release is presented in the next step.

$$P(B) = \begin{cases} [x|x \text{ (delivery performance)} \leq 3\%; \text{to get 10 marks}] \\ [x|x \text{ (delivery performance)} = 4\text{-}10\%; \text{to get 7.5 marks}] \\ [x|x \text{ (delivery performance)} = 11\text{-}20\%; \text{to get 5 marks}] \\ [x|x \text{ (delivery performance)} \geq 20\%; \text{to get 2.5 marks}] \end{cases}$$

The set clearance of back order P(B) shows the performance score of the back-order clearance. It means that the subcontractor cannot achieve the delivery plan but can handle the back-order clearance for the next delivery. The less back-order clearance is the more satisfaction. It is divided into 5 levels; excellent, very good, good, moderate and fails. Each level can get different scores. The 1st level is excellent which holds the back-order between 0 to 3% and get 10 marks. The 2nd, 3rd, 4th level hold the back-order clearance by 4–10%, 11–20%, and over 21% which can get the marks of 7.5, 5, 2.5 respectively.

$$P(Q) = \begin{cases} [x|x \text{ (part defect)} \leq 3\%; \text{to get 10 marks}] \\ [x|x \text{ (NCR clearance)} \leq 3\%; \text{to get 5 marks}] \\ [x|x \text{ (part report inspection)} \geq 95\%; \text{to get 5 marks}] \\ [x|x \text{ (customer claim)} \leq 3\%; \text{to get 5 marks}] \\ [x|x \text{ (quick respond)} \geq 95\%; \text{to get 5 marks}] \end{cases}$$

The set of quality performance shows above that is x when the x is in the various conditions. If the part defect is ≤3%, then it will get 10 marks. If the NCR clearance is ≤3%, then it will get 5 marks. If the part report inspection is ≥95%, it will get 5 marks. If the customer claim is ≤3%, it will get 5 marks. If the quick respond is ≥95%, it will get 5 marks.

The final evaluation for subcontractor evaluation is combined the three criteria; delivery, quality and back order clearance. The total score and the satisfaction level are shown as the followings.

Table 1 shows the total scores and the evaluation criteria. The total score is 80 marks. This is for the monthly evaluation. The remain score is for yearly audit which gives the total score is 30 marks. However, this paper concerns only the monthly evaluation. Presently, the cluster automotive rubber part is evaluated by manual. It is time consuming, high effort, slow problem solving, and cannot deal with uncertainty circumstances and predict future events.

As mentioned from the literature review that fuzzy logic can deal with uncertainty. This paper adopts the fuzzy logic model to deal with the subcontractor performance evaluation. The set representation selects a membership function. Fuzzy set design is a triangular consisting of crisp subsets. The fuzzy linguistic terms are design in 5 levels; very high, high, moderate, low and very low. The conditional fuzzy rules are as the follows.

1. If delivery is low, then performance is low

2. If delivery is moderate, then performance is moderate.

3. If delivery is high, then performance is high.

4. If delivery is very high, then performance is very high.

Criteria	Marks
Delivery	30
Quality	45
Back order clearance	10
Total	80

Score	Grade	Satisfaction level
70–80	A	Excellent
55–69	B	Good
40–54	C	Moderate
< 40	D	To be improved

Table 1.
The score and evaluation criteria.

5. If back order is very low, then performance is very high.

6. If back order is low, then performance is high.

7. If back order is moderate, then performance is moderate.

8. If back order is high, then performance is low.

9. If back order is very high, then performance is very low.

Figure 6 shows the fuzzy logic model for performance evaluation. It consists of two inputs and two outputs. The inputs are delivery and back order whereas the outputs are performance evaluation of the delivery and back order. The model is performed by using MATLAB which is explained later. Fuzzy logic system is an inference engine using for carrying the results based on the input conditions.

Figure 7 shows the fuzzy inference system for delivery performance evaluation consisting of fuzzy input, fuzzification, and fuzzy output. The input membership function is divided into 4 ranges; Very high (VH), High (H), Medium (M), and Low (L). The Mandani method is selected with a triangular form.

Figure 8 shows the fuzzy membership function of the on-time delivery using triangular form. The divided range is referred to the **Table 1** but it needs to adjust a bit in order to fit with the actual form of the supply chain sample model. The low-level ranges are 69.5,75, and 80. The medium-level ranges are 79, 85, and 90. The high-level ranges are 89, 94.92, and 98.92. The very high-level ranges are 98,100,110.3. The total ranges are designed between 60 to 100%.

Figure 9 shows the membership function of the output delivery. The triangle fuzzy model is selected. The Low-level ranges of 5.25, 5.75, and 6. The Medium-level ranges 8.5, 8.75, and 9. The High-level ranges of 11.5, 11.75, and 12. The Very High-level ranges 11.5, 11.75, and 12.

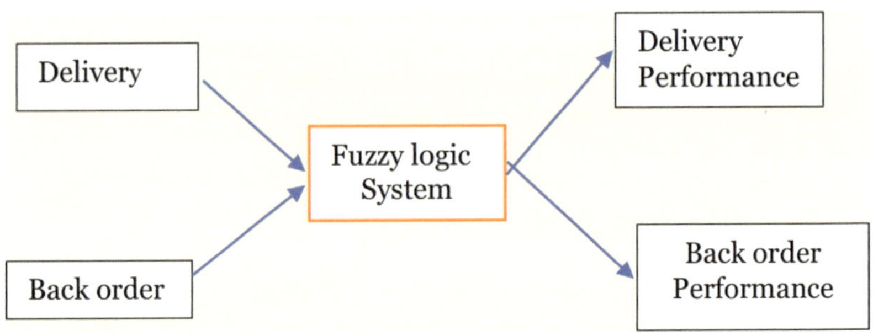

Figure 6.
The fuzzy logic model for performance evaluation.

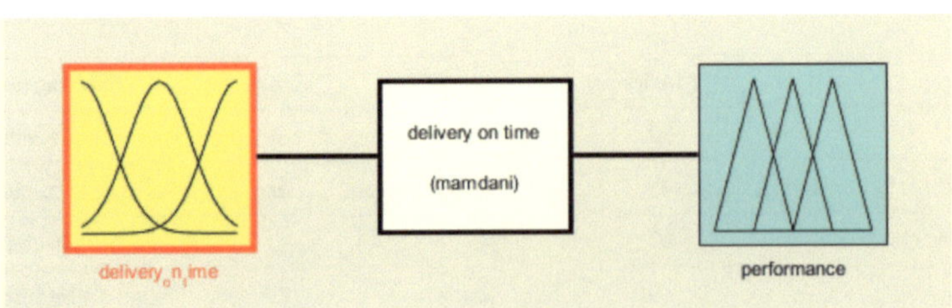

Figure 7.
FIS model for delivery performance.

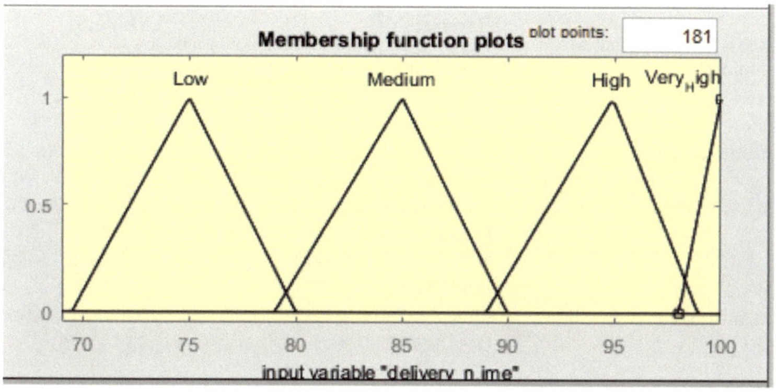

Figure 8.
Membership function input of the delivery.

Figure 10 shows the FIS model for the back-order performance evaluation consisting of fuzzy input, fuzzy inference system and fuzzy output. The model is referred to the **Table 2**. Similarly, **Figure 11** shows the FIS model of the delivery, the back-order clearance membership function which is divided into 4 ranges; Low (L), medium (M), high (H) and very high (VH). The triangular form is selected. The total range is between 0 and 21. The Low range is from 0, 1.5, to 3.99. The Medium range is from 3.95, 7, to 10.99. The High range is from 11, 15, and 20 and the Very High range is from 20.1, 21, to 28.

Figure 12 shows the output membership function, which is divided into 4 levels of Low, Medium, High and Very High. The total range is from 0 to10. The

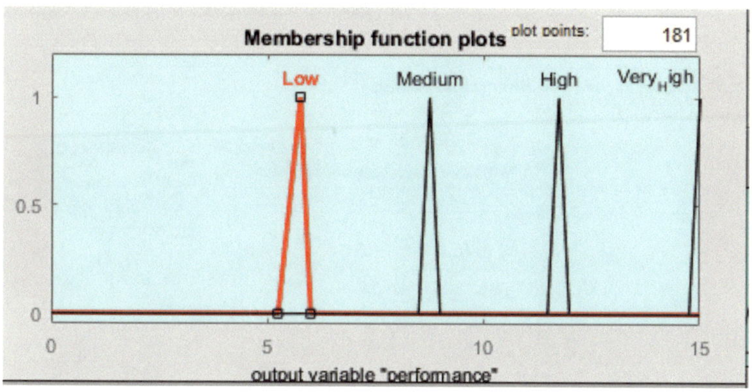

Figure 9.
Membership function output of the performance.

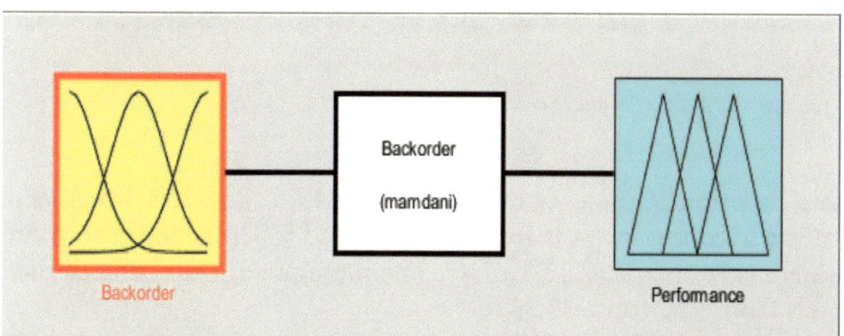

Figure 10.
The FIS model for the back-order performance.

[Input1]	[Input2]	[Input3]
Name = 'input1' Range = [6 15] NumMFs = 4 MF1 = 'in1mf1':'trimf', [3 6 9] MF2 = 'in1mf2':'trimf', [6 9 12] MF3 = 'in1mf3':'trimf', [9 12 15] MF4 = 'in1mf4':'trimf', [12 15 18]	Name = 'input2' Range = [2.5 10] NumMFs = 4 MF1 = 'in2mf1':'trimf', [0 2.5 5] MF2 = 'in2mf2':'trimf', [2.5 5 7.5] MF3 = 'in2mf3':'trimf', [5 7.5 10] MF4 = 'in2mf4':'trimf', [7.5 10 12.5]	Name = 'input3' Range = [1 40] NumMFs = 4 MF1 = 'in3mf1':'trimf', [−12 1 14] MF2 = 'in3mf2':'trimf', [1 14 27] MF3 = 'in3mf3':'trimf', [14 27 40] MF4 = 'in3mf4':'trimf', [27 40 53]

Table 2.
Case of subcontractor neuro-fuzzy performance evaluation.

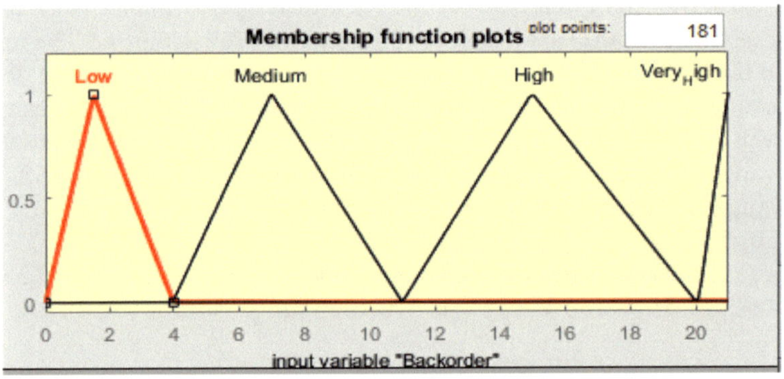

Figure 11.
Membership function input for back order performance.

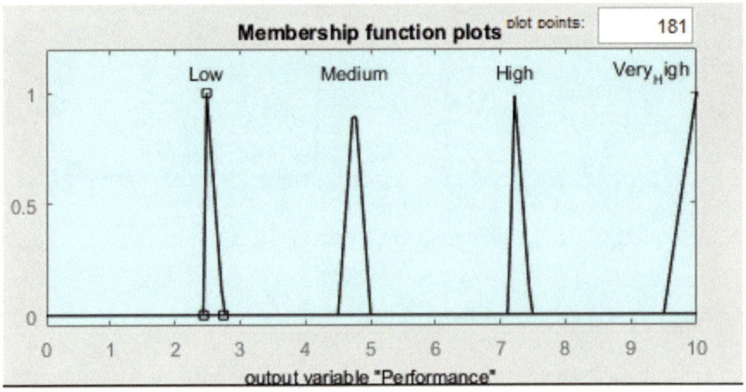

Figure 12.
Membership function output performance evaluation.

membership function of the Low range is from 2.45, 2.5, to 2.75. Membership function of the Medium range is from 4.5,4.75, and 5. The membership function of the High range is from 7.12, 7.22, to 7.47. The membership function of the Very High range is from 9.5, 10, to 13.33.

Figure 13 shows the FIS model is considered and evaluated individually. The ultimate goal of the fuzzy -based system is to get the overall performance evaluation. However, the overall performance has to include quality of the produce part.

Therefore, the model is designed to add quality to be another input membership function and adopt neuro-fuzzy system to solve the problem.

Figure 14 shows the neuro fuzzy inference (ANFIS) system for evaluate overall performance. It consists of 3 inputs; delivery, back order, and quality and one output. The output is overall performance evaluation. The sugeno is selected for the ANFIS rules of inference.

Figure 14 shows the neural training and learning data and the output. **Table 2** shows the overall input membership functions for the ANFIS system. The range represents the minimum and maximum value of the membership function. For example, the range of input 1 (delivery) is 6 to 15. The MF1 (Low range) represents 3 points of the triangle in the base line projection. They are 3, 6, and 9 respectively.

Figure 15 shows input membership functions plots. It consists of 4 levels; low, medium, high and very high. The low ranks between 0 and 2.5. The medium ranks between 2 to 5. The high ranks between 4.75 and 7.5. The very high is fallen over 7.25. The total length is scoped between 0 and 10.

Figure 16 shows the input membership function of the delivery. It is divided into 4 levels; low, medium, high, and very high. The low ranks from 6 to 8. The medium ranks from 8 to 12. The high ranks from 12 to 14 and the very high ranks from over 14. The total length is bounded from 6 to 15.

Figure 17 shows the input membership function plots. It is divided into 4 levels; low, medium, high, and very high. The low starts from 0 to 11. The medium ranks

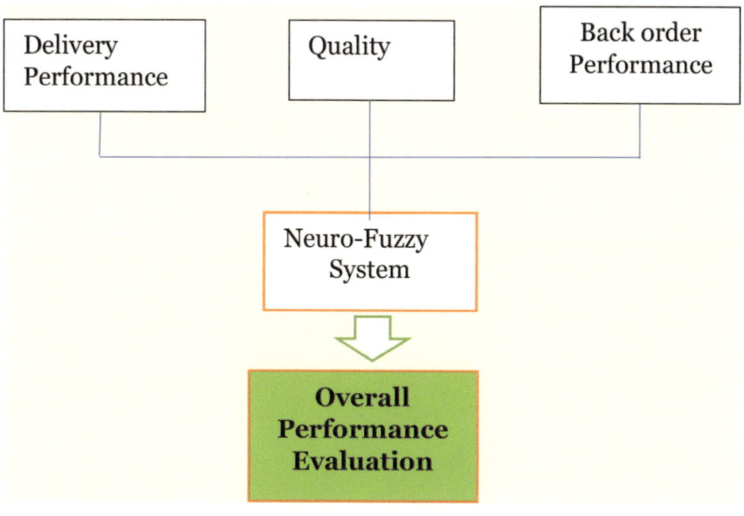

Figure 13.
The neuro-fuzzy model applied for overall performance evaluation.

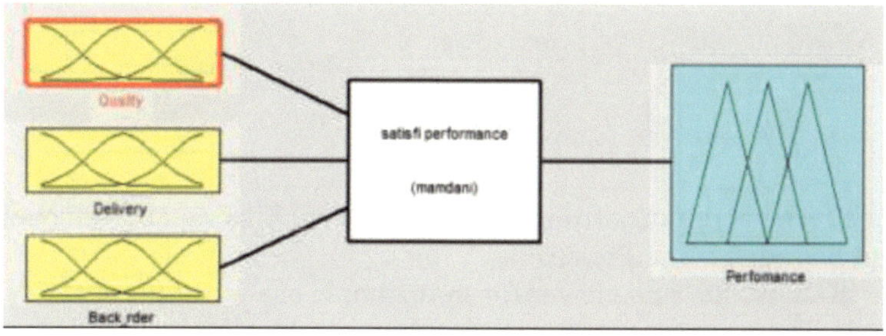

Figure 14.
ANFIS of the neuro-fuzzy performance evaluation model.

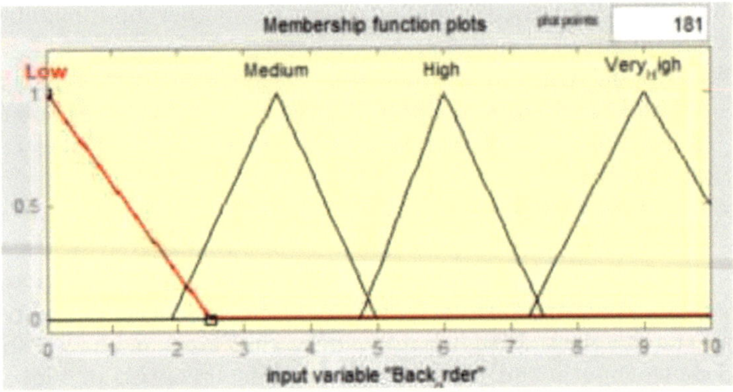

Figure 15.
The input membership function of the back-order clearance.

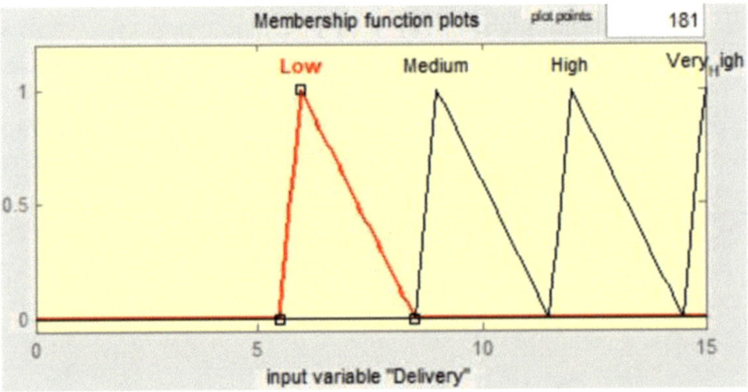

Figure 16.
The input membership function of the delivery.

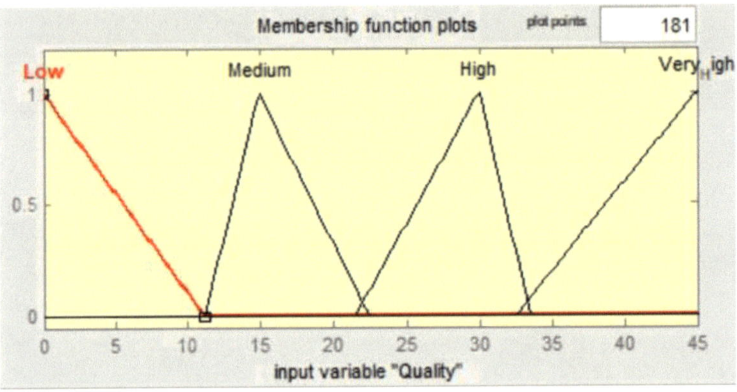

Figure 17.
The input membership function of the quality.

from 11 to 23. The high ranks from 22 to 34. The very high ranks from over 33. The total length bounded from 0 to 45.

Figure 18 shows the rule viewer for the example of the output of the delivery performance. There are four ranges of the input membership function. The example range is in the high which indicates 95.7 and the performance indicates very high. The given score is 11.7 marks in the 15 marks.

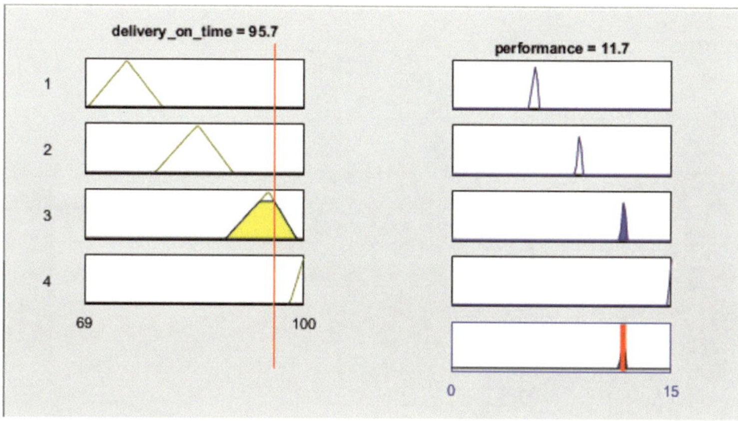

Figure 18.
Rule viewer output performance of the delivery on time.

Figure 19 shows neural training and learning. There are 3 inputs and 1 output. The training is set at 1000 iteration. There are 64 data. The final output is closely 70. The error is 4.546×10^{-6}.

Figure 20 shows the relation surface of the 3 criteria in 3 dimensions. It is found from the relation that the quality is the most influence to the performance whereas the delivery is lower following with the backorder clearance.

Figure 21 shows the rule viewer for the overall performance evaluation model for subcontractor. The data collection from the actual operation. The subcontractor data are inputted into the model and carried out with the result. As the sample above, the score of input 1 (delivery performance) given 15 marks meaning that the delivery performance is very high (99–100%). The 2nd input of the back-order clearance is received 10 marks meaning that this subcontractor performed very well. The percentage of back-order clearance is 1–3%. The 3rd input (input 3) is quality of product and service to the OEM agent and this subcontractor received 40 marks. The quality includes the products delivered quality to the OEM agent.

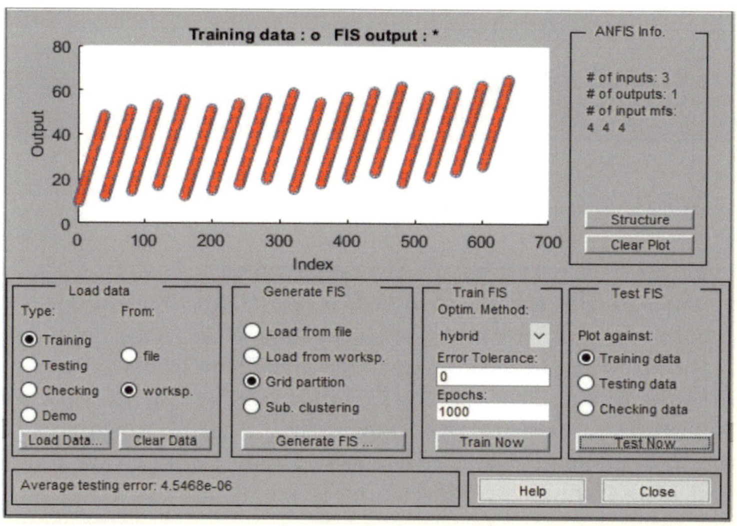

Figure 19.
Neural training and learning.

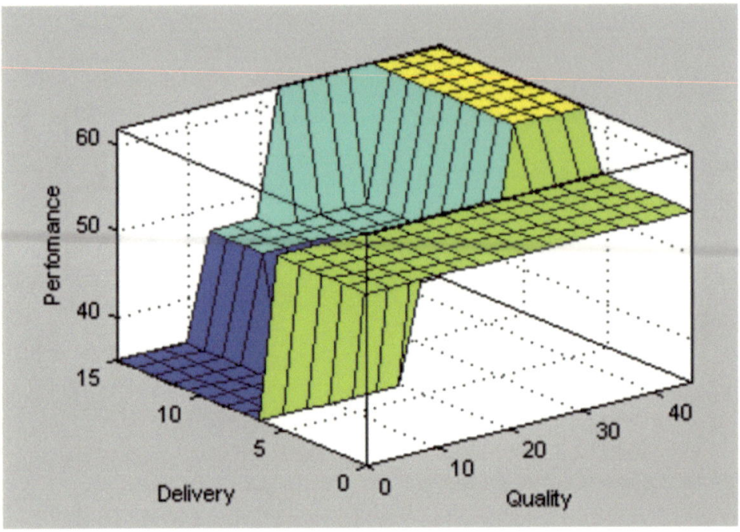

Figure 20.
The relation surface of the 3 criteria in 3 dimensions.

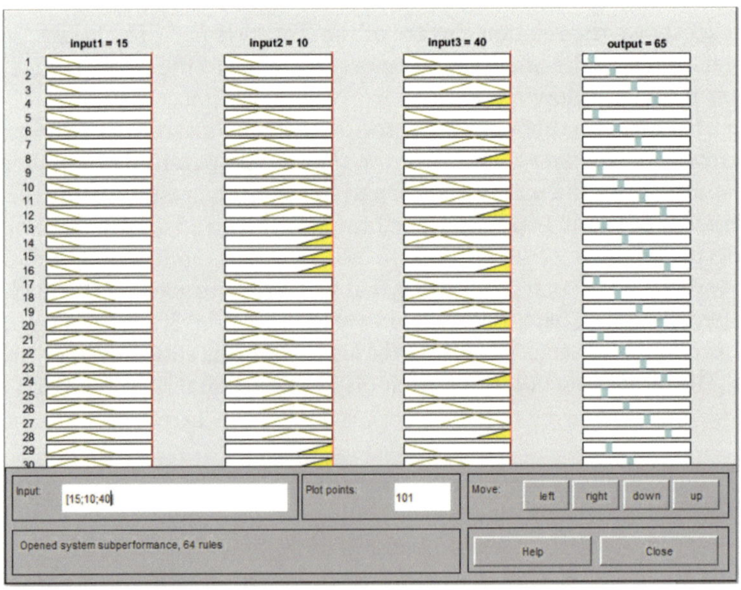

Figure 21.
The rule viewer for the performance evaluation model.

It measures as ppm (part per million). The rank score is A, B, C, D. A is given by the ppm more than 90%. B is more than 80%. C is more than 70% and D is less than 70%. In addition, the quality is measured by defects, NCR (non-conformance record), back order clearance, communication, and claim from customer.

Table 3 shows validation of the ANFIS comparison with actual performance calculation. They are 13 subcontractors are studied (A-K). The column #2 is given marks that calculated by manual. The column #3 and column #4 are also the given marks are calculated by manual. The column #5 and column #6 are determined by the fuzzy inference method. The column #7 and column 8 show the comparison between the actual performance and the neuro fuzzy subcontractor performance. It is found that the neural fuzzy system performs effectively with error 0%.

DOI: http://dx.doi.org/10.5772/intechopen.93679

Subcontractors	Delivery	Back order	Quality	Fuzzy logic delivery	Fuzzy logic back order	Actual performance	Neuro fuzzy subcontractor performance	Error%
A	12	7.5	40	11.7	7.3	59	59.00	0
B	15	10	40	14.9	9.84	64.74	64.74	0
C	15	10	20	14.9	9.84	44.74	44.74	0
D	12	10	40	11.7	9.84	61.54	61.54	0
E	15	10	40	14.9	9.84	64.74	64.74	0
F	15	10	20	14.9	9.84	44.74	44.74	0
G	15	10	35	14.9	9.84	59.74	59.74	0
H	15	10	30	14.9	9.84	54.74	54.74	0
I	15	10	20	14.9	9.84	44.74	44.74	0
J	15	10	20	14.9	9.84	44.74	44.74	0
K	15	10	40	14.9	9.84	64.74	64.74	0
L	15	10	35	14.9	9.84	59.74	59.74	0
M	15	7.5	35	14.9	7.3	57.2	57.20	0

Table 3.
The validation of the ANFIS comparison with actual performance.

Quality	Delivery	Back order	Performance	Quality	Delivery	Back order	Performance
1	1	1	1	11.25	6	2.5	19.75
1	1	2	1	11.25	6	5	22.25
1	1	3	1	11.25	6	7.5	24.75
1	1	4	1	11.25	6	10	27.25
1	2	1	1	11.25	9	2.5	22.75
1	2	2	1	11.25	9	5	25.25
1	2	3	1	11.25	9	7.5	27.75
1	2	4	1	11.25	9	10	30.25
1	3	1	1	11.25	12	2.5	25.75
1	3	2	1	11.25	12	5	28.25
1	3	3	1	11.25	12	7.5	30.75
1	3	4	1	11.25	12	10	33.25
1	4	1	1	11.25	15	2.5	28.75
1	4	2	1	11.25	15	5	31.25
1	4	3	1	11.25	15	7.5	33.75
1	4	4	1	11.25	15	10	36.25
2	1	1	1	22.5	6	2.5	31
2	1	2	1	22.5	6	5	33.5
2	1	3	1	22.5	6	7.5	36
2	1	4	1	22.5	6	10	38.5
2	2	1	1	22.5	9	2.5	34
2	2	2	1	22.5	9	5	36.5
2	2	3	1	22.5	9	7.5	39
2	2	4	2	22.5	9	10	41.5
2	3	1	1	22.5	12	2.5	37
2	3	2	1	22.5	12	5	39.5
2	3	3	2	22.5	12	7.5	42
2	3	4	2	22.5	12	10	44.5
2	4	1	2	22.5	15	2.5	40
2	4	2	2	22.5	15	5	42.5
2	4	3	2	22.5	15	7.5	45
2	4	4	2	22.5	15	10	47.5
3	1	1	2	33.75	6	2.5	42.25
3	1	2	2	33.75	6	5	44.75
3	1	3	2	33.75	6	7.5	47.25
3	1	4	2	33.75	6	10	49.75
3	2	1	2	33.75	9	2.5	45.25
3	2	2	2	33.75	9	5	47.75
3	2	3	2	33.75	9	7.5	50.25
3	2	4	2	33.75	9	10	52.75

Quality	Delivery	Back order	Performance	Quality	Delivery	Back order	Performance
3	3	1	2	33.75	12	2.5	48.25
3	3	2	2	33.75	12	5	50.75
3	3	3	2	33.75	12	7.5	53.25
3	3	4	3	33.75	12	10	55.75
3	4	1	2	33.75	15	2.5	51.25
3	4	2	2	33.75	15	5	53.75
3	4	3	3	33.75	15	7.5	56.25
3	4	4	3	33.75	15	10	58.75
4	1	1	2	45	6	2.5	53.5
4	1	2	3	45	6	5	56
4	1	3	3	45	6	7.5	58.5
4	1	4	2	45	6	10	61
4	2	1	3	45	9	2.5	56.5
4	2	2	3	45	9	5	59
4	2	3	3	45	9	7.5	61.5
4	2	4	3	45	9	10	64
4	3	1	3	45	12	2.5	59.5
4	3	2	3	45	12	5	62
4	3	3	3	45	12	7.5	64.5
4	3	4	3	45	12	10	67
4	4	1	3	45	15	2.5	62.5
4	4	2	3	45	15	5	65
4	4	3	3	45	15	7.5	67.5
4	4	4	4	45	15	10	70

Table 4.
The fuzzy rules of the ANFIS performance prediction.

Table 4 shows the fuzzy rules of the ANFIS performance prediction. The 3 first columns are input and 4th column is output. The column 5–7 are generated by the ANFIS. The last column is the performance prediction. There are 64 cases or rules. If a rule is changed meaning that the performance will be changed. So that, the performance can be predicted.

4.3 Cloud production online monitoring (CPOM) system

This section explains the cloud computing online monitoring system as shown in the **Figure 22**. The CPOM system links the production process and monitor, and evaluation. The production monitoring and control starts at the P&D orders which are received from the 1st tier company every 15 days. It is sometimes that the uncertain orders are given to the subcontractors to increased or decreased quantity of the parts. The production plan is assigned based on the availability of the factory. Capacity is presently checked as well as inventory, back order quantity, and raw material stock on hand. CIM system calculates the availability and actual capacity for creating aggregate planning, master production scheduling, and job release planning. The plans are delivered to shop floor supervisors including daily production plan, machine used plan, and due date delivery plan. Then, production is operated by serial processes. The compression mold or injection mold is the first operation. The 2rd station is de-flashing and finishing the parts. The last station is quality inspection, counting, recording and packing. Instead of doing so by manual, the research paper presents the novel method which employs the CPOM system to increase productivity of both machines and operators as well as the competitiveness. The system is divided into two main parts; the 1st CPOM system and sub CPOM system which is designed as the private cloud under the company security policy. The 1st CPOM is able to monitor and tracking the production process for supporting rapid decision making of the SCM management team. Besides, the critical control of the 1st tier is the raw material mixing company and mold design company which are supporting resource companies and sharing costs as well as quality standard control under specific customer requirement.

4.3.1 Sensor design and QR code reader

Sensor is a heart of the CPOM system. It works as the interface signal processing from objects to the system input. There are many types of the sensors; fiber optic sensor, photo electric sensor, proximity sensor, area sensor, pressure sensor, multi-functional sensor. Such sensors can be used together with precision measurement, PLC, vision system, laser marker, barcode reader, digital microscope and so on. The CPOM system selects fiber optic amplifier. It can easily set light using automatic mode. It is amplified the light around 250 times. The respond speed is 50 μs. The monitor output can be LED or graph. The output can be single, double or analog

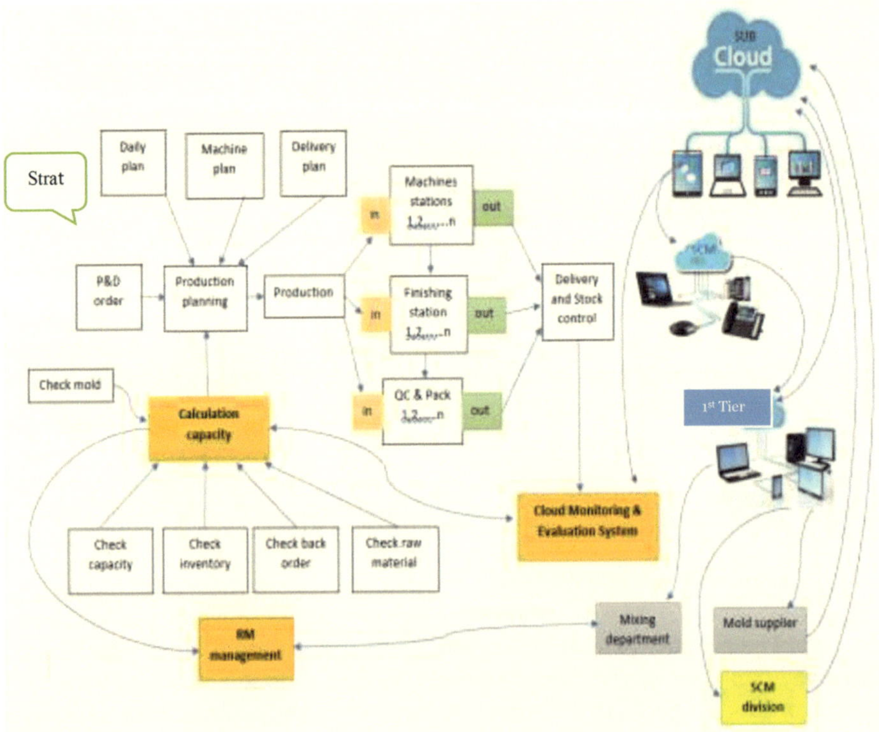

Figure 22.
The CPOM system development.

output. The right **Figure 23** shows the relation of light intensity and displacement. There are different ranges but the effective performance is set on the certain displacement.

QR recorder is used to track a symbolic information of the production process. The QR code generator can be simply done by software and can link to the QR code reader. **Figure 24** shows the QR code reader using a mobile phone.

4.3.2 Conveyor design

The conveyor is designed for delivering the part in the final operation. The conveyor includes sensors, QR code reader, and cloud interface using ESP32. Motor drive is designed for adjustable speed. The displacement between sensor and parts must be effective and eliminate any error.

Figure 25 shows the types of the conveyor system which are used in the factory linked to the cloud management system. They are covered the whole features of the part constraints.

Figure 26 shows the types of the sensor used in the factory. The first sensor type is tracked in the horizontal position while the parts are moved by the conveyor. The second type of the sensor is tracked on the vertical while the parts are delivered through the silo. This is designed and developed for the flat part shape. The third type of the sensor is portable movement which is fit with the large size of the parts. The fourth sensor is called axillary types which is designed to combined with the push button switch and light linked the cloud management system. This type is suitable for the ununiform shape of the parts.

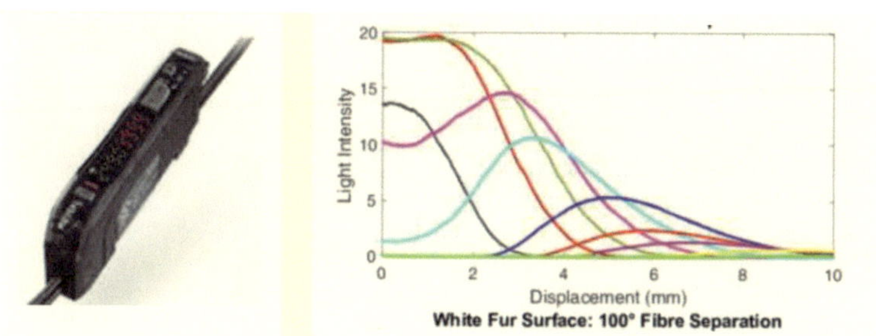

Figure 23.
Fiber optic sensor and characteristic.

Figure 24.
QR code generator and reader.

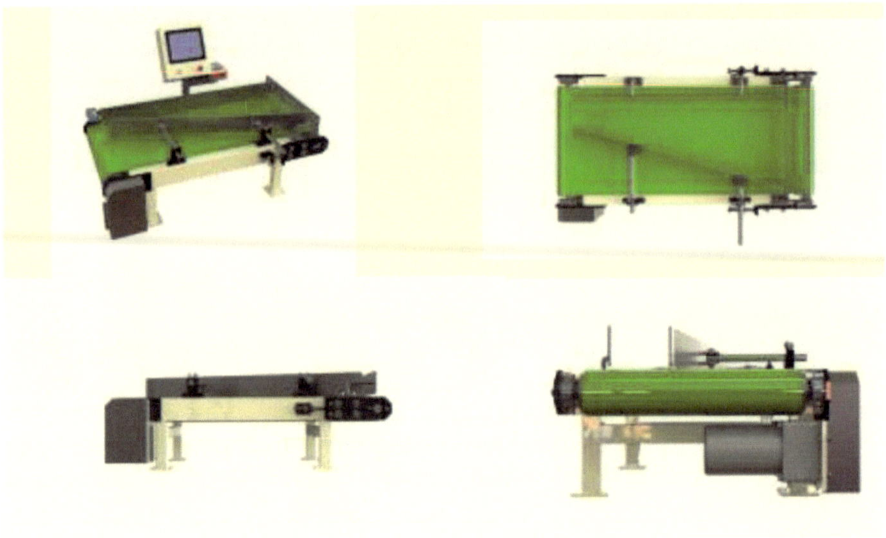

Figure 25.
The conveyor system for delivery finished good.

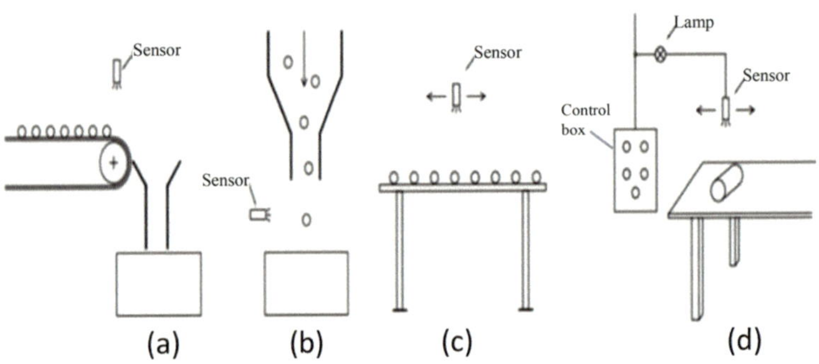

Figure 26.
The types of the sensor used in the factory.

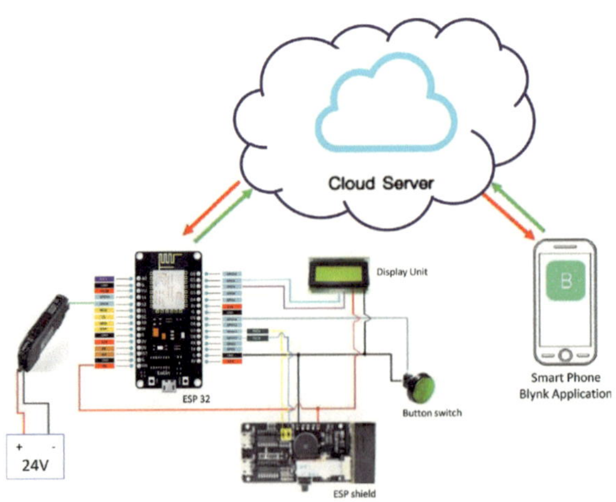

Figure 27.
Cloud interface using ESP32 WiFi.

4.3.3 Cloud interface design

Cloud interface design is used for data tracking and CPOM communication by real-time via WiFi and Bluetooth. It has 18 channels, 3 SPI interface, 3 UART interfaces, 2 12C interfaces, 16 PWM output channels, 2 digitals to analog converter (DAC), 2 12S interfaces, and 10 capacitive sensing GPIOs. **Figure 27** shows the cloud interface using ESP 32 Wifi, Node-Red and blynk application which is linked to a smart phone.

5. Implementation, results and discussion

The section explains the implementation of the CPOM system and shows the result of usages. The example of the P&D order is shown in the **Figure 28** for a month. The delivery date is on the 8, 10,11, 23, 26, and 28. The order # 3 is due on the 8 and parts are completely finished on the 7. It is shown by the green color.

This section explains only one point of the tracking production process system on cloud via WiFi ESP32 micro controller. All production information is tracked and delivered by QR code. The most benefits are to apply the system to the whole supply chain and discuss uncertain situation and collaborative solving problems. This section does not show other dash board for tracking and reporting production progress in every work stations. It is possible to do so via the same CPOM platform with extend the data.

No.	Job No.	Part No.	1	2	3	4	5	6	7	8	9	10	11	12	13	14	15	16	17	18	19	20	21	22	23	24	25	26	27	28	29	30	31	PlanQty	ActualQty	BadOrder
1	PC6306-0001	17505-20V1-81-25										100																						100		
1	PC6306-0001	17505-20V1-81-25																																		100
2	PC6306-0002	17505-20V1-81-25											250																					250		
2	PC6306-0002	17505-20V1-81-25																																		250
3	PC6306-0003	4W4-Q9259								500																								500		
3	PC6306-0003	4W4-Q9259							500																										500	
4	PC6306-0004	6III3-9L00																										450						450		
4	PC6306-0004	6III3-9L00							100																										100	350
5	PC6306-0005	9D54-YH4-6I00																												222				222		
5	PC6306-0005	9D54-YH4-6I00																																		222
6	PC6306-0006	SR-5051																							120									120		
6	PC6306-0006	SR-5051							100																										120	

Figure 28.
The report delivery monitoring on cloud management system.

6. Conclusion

The paper presented automotive industrial supply chain performance evaluation under uncertain constraints on cloud computing system. The supply chain in rubber part industry is explained in details. The new concept of the CPOM system is developed and tested in the factory. It can enhance efficiency of production control and monitoring as well as decision making under the uncertain circumstance. The CPOM is explained from the design to implementation and the results of application.

Author details

Suthep Butdee
Department of Production Engineering, Faculty of Engineering, King Mongkut's University of Technology North Bangkok, Bangkok, Thailand

*Address all correspondence to: suthep.butdee@gmail.com

References

[1] Bowersox DJ, Closs DJ, Cooper MB, Bowersox JC. Supply Chain Logistics Management. 4th ed. Singapore: McGraw-Hill International Edition; 2013

[2] Hopp WJ, Spearman ML. Factory Physics. 2nd ed. Singapore: Irwin McGraw-Hill; 2000

[3] Zadeh LA. Fuzzy sets. Information and Control. 1965;**8**(3):338-353

[4] Szmidt E. Uncertainty and information: Foundations of generalized information theory (Klir, G.J.; 2006). IEEE Transactions on Neural Networks. 2007;**18**(5):1551-1551. DOI: 10.1109/tnn.2007.906888

[5] Chen SG. Fuzzy scored based logistics management in robust SCM. Computers and Industrial Engineering. 2012;**62**:740-745

[6] Ganga GMD, Carpineti LCR. A fuzzy logic approach to supply chain performance management. International Journal of Production Economics. 2011;**134**:177-187

[7] Butdee S, Nitnara C. A fuzzy logic combined with LP model for performance evaluation to distribute purchase orders in cluster manufacturing. Procedia Manufacturing. 2019;**30**:19-25

[8] Zakaria M, Shebany M, Sarahan S. Artificial neural networks: A brief overview. International Journal of Engineering Research and Applications. 2014;**4**(2):7-12

[9] Kocamaz UE, Goksu A. Control and synchronization of chaotic supply chain using intelligent approaches. Computers and Industrial Engineering. 2016;**102**:476-487

[10] Zayegh A, Bassam NA. Neural Network Principles and Applications. UK: IntechOpen – Provision chapter; 2018

[11] Feng S, Li L, Cen L, Huang J. Using MLP networks to design a production scheduling system. Computers and Operations Research. 2013;**30**:821-832

[12] Lee I, Shaw MJ. A neural-net approach to real time flow-shop sequencing. Computers and Industrial Engineering. 2000;**38**:125-147

[13] Che ZH. PSO-based back-propagation artificial neural network for product and mold cost estimation of plastic injection molding. Computers and Industrial Engineering. 2010;**58**:625-637

[14] Efendigil T, Onut S. An intergration methodology based on fuzzy inference systems and neural approaches for multi-stage supply chains. Computers and Industrial Engineering. 2012;**62**:554-569

[15] Chupin MM, Katasev AS, Akhmetvaleev AM, Kataseva DV. Neuro-fuzzy model in supply chain Management for Objects State Assessing. International Journal of Supply Chain Management. 2019;**8**(5):201-208

[16] Didehkhani H, Jassbi J, Pilevan N. Assessing flexibility in supply chain using adaptive neuro fuzzy inference system.Hong Kong, China; IEEE. 2009:513-517. DOI: 10.1109/IEEM.2009.5373292

[17] Sremac S, Zavadskas EK, Matic B, Kopic M, Stevic Z. Neuro-fuzzy inference systems approach to decision support system for economic order quantity. Economic Research. 2019;**32**(1):1114-1137

[18] Tipparat T. Application of Adaptive Neuro Fuzzy Inference Systems in Supply Chain Management Evaluation.

UK: Intech Open Access Book Chapter; 2012. pp. 115-126

[19] Mahmood K, Lanz M, Toivonen V, Otto T. A performance evaluation concept for production systems in an SME network. Procedia CIRP. 2018;**72**: 603-608

[20] Kaganski S, Majak J, Karjust K, Toompalu S. Implementation of key performance indicators selection model as part of the enterprise analysis model. Procedia CIRP. 2017;**63**:283-288

[21] Amrina E, Yusof SM. Key performance indicators for sustainable manufacturing evaluation in automotive companies. Singapore; IEEE. 2011: 1093-1097. DOI: 10.1109/IEEM. 2011.6118084

[22] Hon KKB. Performance and evaluation of manufacturing systems. CIRP Annals. 2002;**54**(2): 139-154

[23] Behrouzi F, Wong KY. A study of lean supply chain performance measure of SMEs in the automotive industry. In: Industrial Engineering and Engineering Management. Singapore: IEEE; 2001. pp. 237-241. DOI: 10.1109/IEEM.2011. 6117914

[24] Roda I, Macchi M. Factory level performance evaluation of buffered multi-state production systems. Journal of Manufacturing Systems. 2019;**50**: 226-235

[25] Joppena R, von Enzbergb S, Gundlach J, Kuhnc A, Dumitrescud R. Key performance indicators in the production of the future. CIRP. 2019; (81):759-764

[26] Hyytia E, Righter R. Simulation and performance evaluation of mission criteria dispatching systems. Performance Evaluation. 2019; **2019**(135):102038

[27] Jagan MR, Neelakanteswara K, Krishnanand RA. A review on supply chain performance measurement systems. Procedia Manufacturing. 2019; **30**:40-47

[28] Hudson M, Smart A, Bourne M. Theory and practice in SME performance measurement systems. International Journal of Operations & Production Management. 2001;**21**(8): 1096-1115

[29] Syalina I, Simunovic G, Saric T, Lujic R. Evolutionary neuro fuzzy systems for surface roughness evaluation. Applied Soft Computing. 2017;**52**:593-604

[30] Zhang L, Li BH. Cloud manufacturing: A new manufacturing paradigm. Enterprise Information Systems. 2010;**8**(2):167, 187

[31] Wang P, Fan ZA. Cloud computing for cloud manufacturing: Benefits and limited. Journal of Manufacturing Science and Engineering. 2015;**137**:1-6. DOI: 10.1115/1.4030209

[32] Available from: www.comparethecloud.net/article/the-role-of-ai-in-powering-enterprise-content/

Chapter 14

An Investigation of the Metal Additive Manufacturing Issues and Perspective for Solutions Approach

Omar Ahmed Al-Shebeeb

Abstract

Metal Additive Manufacturing (MAM) is delivering a new revolution in producing three-dimensional parts from metal-based material. MAM can fabricate metallic parts with complex geometry. However, this type of Additive Manufacturing (AM) is also impacted by several issues, challenges, and defects, which influence product quality and process sustainability. In this chapter, a review has been made on the types of small to medium-sized metallic parts currently manufactured using the MAM method. Then, investigation was undertaken to analyze the defects, challenges, and issues inherent to the design for additive manufacturing, by using MAM method. MAM-related obstacles are discussed in depth in this chapter and these obstacles occur in all size of metal printed parts. The reasons and solutions presented by previous researchers of these obstacles are discussed as well. A potential approach based on the author's knowledge and analysis for solving these issues and challenges is suggested in this chapter. Based on the author's conclusion, the MAM is not limited by part size, material, or geometry. In order to validate the potential solutions developed by the author of this work, performing actual MAM process is required and a local visit to manufacturing factories are also important to visualize these challenges and issues.

Keywords: metal additive manufacturing, powder bed fusion, direct metal deposition, MAM alloys, MAM problems, 3D printing

1. Introduction

The manufacturing field has developed tremendously in the last decades. Today, transforming raw materials into useful parts or products (by using different manufacturing processes) is nearly unlimited. The manufacturing research is always focusing on developing and using manufacturing processes with lower cost and waste, and higher production rate. Currently, there is fast and high variation in the design of products, and very high market competition because of global competition. Design for Additive Manufacturing (DFAM) is one of the important methods for obtaining this global competition in terms of manufacturing.

Additive Manufacturing (AM), is a type of printing (3D printing) method used to create three-dimensional products by laying the fused material layer by layer.

AM is the current and dominant future manufacturing method [1]. AM processes are considered easier as compare to subtractive processes represented by machining and other manufacturing types. This is because of producing a part through one AM process is more effortless than producing the same part through several subtractive manufacturing processes (such casting then machining). Subtractive manufacturing processes often require millions of dollars, while using AM processes can offer the same manufactured parts at a fraction of the cost, and in less than half the time [2]. In addition, manufacturing the part in one process eliminates the need for several skilled workers (which subtractive manufacturing requires), in lieu of a single knowledgeable worker.

AM is the future face of the industry, not just in manufacturing field—printing technology has even been used to construct buildings in recent years. However, the MAM process for producing small and medium size metallic parts also present difficulties and issues. These difficulties may be inherent to the MAM process itself, such as selecting the right MAM process, adjusting support materials, building direction, geometry (complexity) of the part, and printing orientation. All these difficulties might lead to some issues and defects in MAM products, such as porosity, variation in mechanical properties, microstructure evolution, residual stress, fatigue, and crystalline phase and more. In the next section a recent comprehensive literature review is presented to explain the main issues and difficulties facing MAM processes, in the production of small and medium size parts.

2. Literature review

Additive Manufacturing (AM) was employed for the first time in 1987, through use of a stereolithography (SLA) system to build the first three-dimensional model (solidifying a thin light-sensitive layers liquid polymer by using ultraviolet [UV] rays) [3]. Starting in 2017, manufacturing companies focus began orienting manufacturing processes in a new direction; these companies started to depend on AM methods to produce parts. AM methods have grown over the last year, becoming the key area of interest for many such companies [4]. However, the main focus of this literature review related to the MAM method for producing small and medium metallic parts—and asking what the main issues and challenges facing this process are.

This concerns frameworks for selecting the right MAM process, adjusting support material, building direction, geometry (complexity) of the part, and printing orientation. Some of these challenges can interrupt the process itself and cause other issues (such as defects in the final manufactured product, including: porosity, variation in mechanical properties, microstructure evolution, residual stress, and crystalline phase heat accumulative and thermal behavior). Inspection difficulties in MAM have been investigated, with the conclusion that inspection ability and metallurgical validation of highly optimized shape must be integrated with product design and process parameter during and after the CAD design [5]. Several challenges facing MAM methods have been investigated, such as financial consideration, certification and regulation, repeatability, and skills gap. Some suggestions have been made to eliminate these challenges and this researcher discusses further potential solutions for these challenges [6]. Serena et al., studied the difficulties and challenges to be facing when designing for metal additive manufacturing in Selective Laser Melting Additive Manufacturing (SLM AM) process for producing professional sport equipment (medium size cam component). These difficulties, such as functionality, manufacturability, assembly, and printability. All of these difficulties should be considered in the designing of metal additive manufacturing. They present many suggestions to enhance the redesign and printability the cam system [7].

Chergui et al. studied the schedule problem and nesting in production of MAM method by developing mathematical model in Python with a heuristic approach. The heuristic approach has been explained step-by-step as a result, improving the scheduling and nesting problems [8]. Bradley et al. created a seven degree-of-freedom, dual-arm hydraulic by using the SLM method for subsea use. Researchers have described the hydraulic system and how they created the titanium manipulators for Naval research. They have also discussed lessons learned throughout the project and process (design) changes for the future [9]. For particle size, heat treatment, shape analysis, and hardness testing, microstructural analysis in MAM, several solutions have been presented, discussed and applied [10].

Li et al. have discussed the residual stress issue in printing metal parts due to rapid heating rate, rapid solidification with high cooling, and melting back the layers that previously melted. They used Powder Bed Fusion (PBF) and Direct Energy Deposition (DED). This work was performed on small and complex geometry parts of nickel based super alloys, titanium alloys, and stainless. The results showed there is a connection between the residual stress and microstructure: the residual stress has been measured in both as-build metal part and post-processed [11]. Reza Molaei and Ali Fatemi have reviewed the main factors that influence fatigue behavior in producing metal parts. In this review, they have collected some multiaxial fatigue data for Selective Laser Melting (SLM), which is based on Powder Bed Fusion (PBF). They used titanium alloy (Ti-6Al-4 V) as metal powder for printing small sized parts [1].

In terms of energy consumption in MAM, and how it affects the quality of parts produced by MAM methods, ZY Liu et al. have studied the effect of energy consumption on microstructure and mechanical properties. Researchers have also used different types of metallic parts to perform investigations on both the machine and the process levels. On the machine level, they have studied the high-energy tool, control system, and cooling system whereas on the process consumption level, they have investigated energy flow distribution [12]. Bintao et al. reviewed the defects occurred in Wire Arc Additive Manufacturing (WAAM) related to microstructure and mechanical properties (deformation, porosity, and cracking) for high scaled fabricated components and high deposition rate. They used different size of metallic parts for different alloy types (Titanium Alloy, Aluminum Alloy and steel, Nickel based super alloy, and other alloys). Researchers concluded that WAAM is still facing many challenges in producing different materials. The WAAM needs to perform sustainable system in a reasonable time frame. There is a need to produce defect free products by using WAAM. Finally, they suggested ways to improve quality of the product [13]. Filippo et al. developed a methodology for cooling system (impinging air jet) in WAAM process to prevent heat accumulation which increases the workpiece average temperature and consequently affects the WP quality. They used small parts of Fe alloys in their study. Their results showed that the impinging air jet can prevent the heat accumulation on part produced by WAAM [14]. Bintao et al. used in-situ temperature measurement method to analyze the heat accumulation and thermal behavior in Gas Tungsten Wire Arc Additive Manufacturing (GT-WAAM) process. They used medium sized titanium alloy (Ti6Al4V) parts. The result shows that the microstructural morphology, crystalline phase, mechanical properties and fracture feature have been changed when the heat accumulation was along the building direction [15].

Xuxiao Li and Wenda Tan built a numerical model to investigate the three-dimensional grain structure in Direct Laser Deposition (DLD) for stainless steel 304 material. They enhanced the three-dimensional grain structure by using nucleation mechanism. The investigation showed that the nucleation mechanism used in this research could play a role in modifying grain structure in MAM method

[16]. Jun Du et al. developed and tested newly proposed AM method based on Metal Fused Coating Additive Manufacturing (MFCAM). This method is a combination of Fused Metal Coating and Laser Surface melting (bed-based process). They used small parts of 7075 aluminum alloy to prove the experimental work [17]. Christoph et al. developed a computational model to study the critical influence of powder cohesiveness on powder recoating process. Researchers focus on the relationship between the powder particle size and powder layer quality. Small parts of Ti-6Al-4 V were used in this study. As results from this study, decreasing the particle size (increase cohesiveness) will decrease the powder layer quality with highly non-uniform surface profile. In addition, the particle size plays the main role in mechanical properties of powder layer [18].

Lawrence et al. reviewed the development of droplet 3D printing (Droplet Additive Manufacturing). This process was used in producing large and small sized parts for three decades. They discussed the issues regarding process optimization, product structure and properties influenced by oxidation. Their investigation ended up with the conclusion that using the Droplet 3D printing process can change the structure of product, thereby reducing the weight, cost, and increasing strength [19]. Mercado et al. studied the stability and microstructure of large sized parts of nickel-base metal matrix produced by building Plasma Transferred Arc Additive Manufacturing (PTAAM) system; this process can build high scale 3D printed parts. Their study concluded that the PTAAM system has capability to build 3D printed part on nickel-base metal matrix with tungsten carbide wear resistance [20]. Yoozbashizadeh et al. developed new Novel AM method to fabricate medium sized bronze-aluminum parts with Ceramic. This process has been performed by combining Thermal Decomposition for Salt (TDS) method with Powder Bed AM (PBAM) to produce Metal Matrix Composite (MMC). Ceramic particles have been created from TDS, and then combined with bronze-aluminum to create MMC. This process is qualified for aerospace applications [21].

Livescu et al. faced challenges of AM tantalum represented by high melting temperature via utilizing Direct Metal Laser Sintering (DMLS) method. Deposition parameters such as deposition speed and building direction have been analyzed as significant factors to influence on Grain morphology, grain size, crystallographic, and deposition porosity. The authors' results showed that the obtained structure was columnar along the building direction. The deposition condition (speed) has significant effect on microstructural variation. The strip width has the main influence on grain growth [22]. Thao Le et al. tried to combine additive and subtractive strategies to manufacture new part (final part) from end of life part (existing part) by using several additive and subtractive manufacturing processes. They obtained good mechanical properties in final parts. The methodology of combining additive and subtractive manufacturing can be applied by generating process plan for both of them [23].

3. Methods used to create small and medium parts in MAM by factories

Additive Manufacturing (AM) technology, in general, qualifies to build fully functional part in one process without the need for metal removal process which would waste significant amount of raw material. In addition, this process can give more flexibility to the designer in building complex geometry part. Compared to the AM method, the Traditional Manufacturing Technology (TMT) (subtractive manufacturing) has several manufacturing processes. The first level mostly deals with creating the stock material (raw material), and the next level is responsible for material removal process which usually includes several manufacturing processes to obtain the final parts. Currently, the diversification of AM method can be used with variety of materials; it is possible to manufacture metallic parts with high quality and complex geometry by direct and indirect AM method.

3.1 Powder bed fusion meal additive manufacturing (PBF-MAM) process

There are two main categories in MAM method: Powder Bed Fusion (PBF), which is also known as Layer Based Metallic Additive Manufacturing, and Direct Metal Deposition (DMD) [24]. PBF method is based on inert atmosphere or partial vacuum, and it produces parts layer by layer with thickness of 15–20 μm. The energy source for this MAM category is laser or electron beam which is used to fuse and bind the material (powder) on each layer. Whenever the binding of a layer will finish, the table will move down, and new layer of powder will be poured on the previous one and so forth until the part is completed. For preventing the molten (sintered) between one layer and another in substrate, there is a need to use support material which can be made in the pre-processing phase from the same material [24].

3.1.1 Selective laser sintering

The Selective Laser Sintering (SLS) was the first powder-bed based AM process. This process was invented by Ross Householder in 1979 and the first material used in it was amorphous polymer powder or semi-crystalline [3]. The power source in this process was laser, and it was used to sinter (bind) the powder material. First, the model of the printed part needs to be defined by CAD design, the thickness of the layer needs to be specified, and the resulting model will be transferred to the machine software to orient the CAD model in the 3D printer software. The MAM process starts by aiming the laser on a profile representing the shape of the first layer as described by the CAD model from the base of the printed part. The laser power binds the powder layer by layer to create the final part. The new materials which could be produced by this process include ceramic and metallic alloys. The last step in this process is placing the printed part in an oven to remove the polymer (used to bind the particles), sinter the part, and improve the support material. Since this process has high production rate, it is used for rapid manufacturing or prototyping [24, 25].

3.1.2 Direct metal laser sintering

Direct Metal Laser Sintering (DMLS) process was developed by Electro Optical Systems (EOS) in the 1990s. It was used purely for building metallic parts without using the polymer to bind the particles. High power laser was used to melt the metal powder in two dimensions layer by layer. This process offers printed parts with complex shape and geometry with a reasonable cost. The DMLS process utilizes a variety of metals and alloys and operates on the same concept of SLS AM process. The residual stress in product parts is an issue with this process and will be discussed in Section 6.2.1 of this chapter [26].

3.1.3 Selective laser melting

The Selective Laser Melting (SLM) additive manufacturing process is designed to fully melt the metallic powder by using high power-density laser. Therefore, it is considered more powerful and produces parts with better quality and less porosities than either SLS or DMLS processes. At the same time, it works on the same powder-bed concept of both techniques. Because of high temperature of the laser source, shrinkage and thermal distortion are likely to influence the printed parts [27].

3.1.4 Electron beam melting

The Electronic Beam Melting (EBM) is also a powder-bed additive manufacturing process. The difference between EBM and other powder bed AM (SLS, DMLS, and SLM) processes is the source of energy; EBM utilizes electron beam instead of laser. This process was invented by Arcam in 1997. The electronic beam provides higher energy density than laser. This makes the EBM process more flexible with metallic alloys. Nevertheless, this has a drawback of increasing the chances of shrinkage. The EBM has a high scanning speed (several kilometers per second) which helps to reduce heat accumulation between the layers and improves quality [24].

3.2 Direct metal deposition additive manufacturing

Direct Metal Deposition (DMD) additive manufacturing method was invented by Precision Optical Manufacturing (POM). The metallic powder or wire used in the DMD is sprayed directly onto the laser beam or the electric arc beam. The beam melts the metallic powder or wire by the laser or electric arc heat source, then the molten metal drops on bed to build the part layer by layer. Because of that, it is called direct deposition process and not powder bed in DMD. There is a possibility to build multilateral parts from various metallic alloys. To produce complex geometry parts, the nozzle is placed onto a 5-axis CNC machine to offer more flexibility in the movement of the nozzle. Several factors control this method of MAM: layer thickness, deposition rate, nozzle feed rate, laser power, gas flow, and the gap between the nozzle and printing surface [3]. The DMD can be divided into several processes; the next sub-section outlines the main processes of DMD.

3.2.1 Wire arc additive manufacturing

The Wire Arc Additive Manufacturing (WAAM) is used to build large components of titanium, aluminum, steel, and other metals. In this method, arc welding tool and wire (feedstock) are required to perform the process. This process is distinguished by high deposition rate, low material and equipment cost, variety of used materials, and good structure. These features make WAAM process dominant over the other current methods of manufacturing [28].

3.2.2 Gas tungsten-wire arc additive manufacturing

Gas Tungsten-Wire Arc Additive Manufacturing (GT-WAAM) is the same as WAAM process with one difference; a localized gas tungsten shielding is used in GT-WAAM. The tungsten reduces the oxidation on the surface and increases the quality of the layer [15]. There are methods of Direct Energy Deposition (DED) available, and they function on the same concept of WAAM and GT-WAAM (welding method) and they are as follows:

1. Plasma Arc (PA)

2. Gas Metal Arc (GMA)

3. Plasma Transferred Arc (PTA)

4. Laser Beam (LB)

5. Electron Beam Freeform (EBF)

4. Metal additive manufacturing materials (alloys)

Several metals (alloys) can be utilized in the MAM method. In this section, the most important metals (alloys) are listed as follows:

1. **Aluminum alloys:** Al-Si, Al-Si-Mg, and AlSi10 are the most commonly used aluminum alloys for the MAM method. These materials can be used for powder-bed fusion especially in DMLS and SLM process. Performance of these alloys can be enhanced by adding Zr and Sc. Wire based additive manufacturing also uses the aluminum alloys. Aluminum powder is good for parts with thermal properties and low weight, and it is a relatively better choice for parts with thin wall and complex geometry. According to current studies, the aluminum alloys will be the major metal in MAM [29].

2. **Titanium alloys:** Ti6Al47 is the dominant titanium alloy; it can be used for the MAM method. This alloy is mainly implemented with SLM, EBM powder-bed and other MAM processes. However, printing parts with these alloys exhibits residual stress and fatigue as defects which affects the quality of the parts. Ti64 is also assigned as a common titanium alloy for MAM. It has high corrosion resistance and is implemented with DMLS-MAM process. The part produced by this process can be used for medical application, aerospace, firearm, and automotive parts which require high corrosion resistance [9, 15].

3. **Iron-based alloys:** Different types of iron alloys can be utilized in MAM method: 316 L stainless steel, 314 L stainless steel, 15–5 stainless steel, 17–4 stainless steel, and Fe-6.9% Si, maraging steel MS1 and more. These iron-based alloys can be used in powder-bed fusion methods, such as SLM and DMLS. The MAM parts produced from stainless steel can be used for high corrosion resistance and strength applications such as medical, firearms, energy and automation, and tooling applications [30, 31].

4. **Super nickel/chromium alloys:** Nickel/Chromium alloys include IN718, IN625, HX, MK500, and Haynes 282. The parts produced from the powder of these alloys have excellent heat and corrosion resistance with hardness of 40–47 HRC. Different sized parts can be produced from these alloys with different applications such as rockets, space, aerospace, firearm, energy, and automotive. DMLS and SLM MAM methods can also be used with these alloys [11, 30].

5. **Other alloys:** The above-mentioned alloys are the most famous alloys which are used to produce parts by MAM method. However, there are other types of metals that can be utilized to produce parts by different processes of MAM method. Some of these metals can be a real challenge to be melted layer by layer from their powder. For example, Inconel 718 is utilized by EBM process, Inconel 625 is produced by SLM process, copper is printed by EBM process, and Fe65Cu17.5Ni17.5 is printed by SLM. Most of these metals (alloys) can be used to produce parts for different applications, however some of them are still in the stage of study [29].

5. Parts size produced by MAM

MAM method has no limitation on size, shape and geometric complexity. Small, medium and large sized parts with complex geometry can be obtained from MAM method. For example, different size and geometry of gears, pistons, pullies, junctions, turbines, manipulator arm, robot arm, fins, rocket, firearms, automobile parts, impellers, fans, and medical tools. However, printing these parts by the MAM method without difficulties and issues is a challenge. This will be discussed in the next section.

6. Metal additive manufacturing (MAM) challenges, issues, and approached solutions

Metal Additive Manufacturing (MAM) is a process used to fabricate and repair parts that have a complex geometry or need to be functionally graded. The process can be performed by depositing multiple layers to produce the required part. Several MAM methods can be used to manufacture metallic parts. They are categorized into two types according to their method of performing (Direct Metal Deposition and Powder Bed Fusion). A successful manufacturing process depends on the deposition technique used, the parameters selected, and the materials used. Moreover, the MAM depends on deposition conditions such as temperature and protective atmosphere to determine whether cracking and oxidation of the deposited layers occurred or not. In the present study, the MAM methods are discussed and evaluated for different manufacturing parameters and deposition conditions for different metallic materials. Multiple challenges and issues encountering the MAM methods will be discussed in the next sub-sections. This study has shown that some suggested solutions can be successfully processed by MAM by using different methods. A significant number of MAM methods are being utilized these days.

6.1 MAM challenges, issues, and approached solutions

Despite the ability of the MAM method in reducing the complexity of parts to simple 2D slices and machining super hard alloys, the MAM method is still suffering from several issues and challenges which are interrupting its progress. In this sub-section, these challenges and issues will be discussed, and the solution approached by the author will be presented.

6.1.1 Lack of knowledge of design for additive manufacturing

Understanding how the materials work in MAM and how the metals are printed layer by layer to print three dimensional models are big concerns in the design for additive manufacturing. Lack of understanding the differences between plastic 3D printing and metal 3D printing can lead to parts with low quality, high defects, and high probability of failure. Skills gap in MAM method is considered an important issue. The knowledge of transferring several subtractive manufacturing processes to one metal additive manufacturing process is a big step. Selecting the appropriate metals (powder or wire) to print the parts and the appropriate process requires a significant amount of knowledge. Also, understanding the orientation of parts and adjusting the support materials in right place by using the printer software before printing is important to obtain desirable quality and characteristics. Similarly, printing high quality parts with complex geometry is also a big challenge with the MAM method, especially when printing aerospace or firearm parts. Therefore, lack of education issue can cause companies to lag or fail in performing successful MAM.

6.1.1.1 Proposed solution

Well-trained workers or engineers and skilled workforce are necessary to diminish this problem. Building and finding capable workforce is not easy and requires effort and time. It also requires a significant work to shift into new, important method such as MAM. To enable the transition between general AM and MAM, more focus is needed on hiring well educated people in this field. Transition process from conventional manufacturing to MAM does need to shrink the workforce. Small workforce is required in MAM. For example, traditional manufacturing system needs 12 skilled workers and 7 engineers to run and manage the system, but MAM system might only require two well-trained, knowledgeable engineers to perform the same work. Building new techniques such as generative design and generative orientation is also beneficial to perform this transition. Finally, in the opinion of author, transformation to MAM should be executed by an educated and a knowledgeable workforce. This could be obtained by creating a custom plan for every company about the transmission and the risk of it.

6.1.2 Repeatability

The MAM method is still facing a lot of variation in parts. The problem that companies encounter is to create two similar parts, then make these parts repeatedly. America Makes Company is still suffering from this problem in MAM and only 30% from the produced parts meet the specifications of the company. In addition to this problem, the MAM method is also suffering from mass production process [2].

6.1.2.1 Proposed solution

The repeatability inside the MAM system can be acquired by developing the standardization of quality. A standardization should be set for the MAM process. This standardization should be developed after a massive study on every MAM process and from this study several factors, such as MAM process type, printing time, environment, used materials, part size, production rate, complexity of the part, functionality and process setting need to be discussed and assigned to standardize the process. All these factors need to be organized and assigned inside specific range which can provide the quality that is required to obtain the repeatability.

6.1.3 Selecting an appropriate MAM process

Every MAM process has different strength and ability to produce specific metallic part, whether the process is Powder Bed Fusion (PBF) or Direct Energy Deposition (DED) process. Using an appropriate process in MAM method will lead to produce defect-free parts. For example, using EBM process that has a high-power source will allow to produce a thicker layer, and will lead to high production rate. However, using the EBM process to produce tiny and complex part with lot of features will be a challenge because the residual powder in each layer will be semi sintered with thicker layer that will create part with undesired quality. Therefore, complex parts can be produced successfully on a SLM or DMLS system.

6.1.3.1 Proposed solution

Making MAM process selection map is a good solution for eliminating this obstacle. This map will contain inputs and outputs. The inputs will have process parameters, such as part size, part geometry, material used, quality and mechanical properties required, and design parameters. These inputs will be analyzed and addressed, then they will be matched with each MAM process capability. Finally, the outputs will be extracted from all of that and the priority of using the MAM processes will be ordered according to their fitness for printing the required parts.

6.1.4 Materials issues

Using the appropriate material (metal) for MAM is also a challenge in this process. The metals used for printing parts should match the selected 3D printing machine. MAM still faces problems in printing some metals, such as Inconel 718, Inconel 625, titanium, and tantalum. Secondly, some parts have multi-features in their design resulting in a variable mechanical properties for every feature. This will increase the probability of failure for these parts. However, it is worth mentioning that the list of metals can be used in MAM is still relatively short.

6.1.4.1 Proposed solution

Using multi-material (metal) machine will solve the problem of parts with multi features. These machines can provide appropriate material for every feature in complex parts. As a result, it will fix the issue of failure in these parts. Performing this solution is not easy and needs a lot of study to get the machine structure that will be used for printing multi-materials (metals). Especially, there is one important condition; the multi-materials can be banded together to create the final parts. This solution can perform with direct deposition MAM method because the multi-material can be melted separately and then combined, but it will be a big challenge with the powder-bed MAM method.

6.1.5 Heat treatment

The metal 3D printing is different than other types of 3D printing. In metal 3D printing, the printed metal part needs heat treatments to obtain the desired mechanical properties which meets the quality of printed parts. Since the MAM is shifting from rapid prototyping to the actual production: aerospace, firearm applications, highly customized parts, the heat treatments play a significant role in the cost and the quality of the part. Mixing different types of raw materials and creating alloys for machining, such as titanium steel, aluminum is not easy, but it can be done in MAM. However, the parts produced by MAM need heat treatments to obtain the desired mechanical properties.

6.1.5.1 Proposed solution

Producing metal part in MAM method without internal stress is an issue in additive manufacturing. Some MAM machines have heat treatments in their structure. Some produced parts in MAM require more levels of heat treatment to obtain specific mechanical properties, such as the elongation to failure and the fatigue strength. Excellent vacuum furnace is required to perform the heat treatment which eliminates the oxygen from entering the machine and obtain the required temperature control and deep vacuum level. Another issue with this process might be the cost of heat treatment itself. Producing vacuum furnace with these characteristics may cost $100 K, thus the cost of heat treatment part of titanium may reach $600. This solution of using high maintained vacuum furnace might give high corrosion resistance with excellent strength, toughness, and low weight part of titanium alloy. Therefore, this process should be used in producing parts having important and expensive applications.

6.1.6 Financial considerations

The cost of MAM method is an issue as compared to the plastic AM. The cost of some MAM machine may reach up to $350 k. The cost of facility and raw materials (powder or wire) is also expensive. Providing the metallic powder with specific particles size for expensive metals will also add capital expenditures. In addition, with all these expenses, failures in printing the parts will push such MAM companies out of business.

6.1.6.1 Proposed solution

A comprehensive study regarding this financial consideration should be performed. This study will be about the investment and profit that can be obtained with transition to MAM from traditional (subtractive) manufacturing processes. A criterion should be considered about the expenses and cost of MAM and compared to the capital expenditures of subtractive manufacturing. The entire cost for structure should also be noted. Moreover, an evaluation on the throughput and type of the parts being printed should be used to make a final decision about the transition from a subtractive manufacturing process to MAM. This decision should be made between company management because it will affect the entire company's structure including the supply chain.

6.1.7 Certification and regulation

The key of success for MAM is to obtain the quality and reliability required on the printed parts. Certification and regulation means assigning a customization and individualization to the additive manufactured part [2]. Certification process is considered as another challenge which manufacturing companies are suffering from in the MAM field. The MAM method without certification is a disadvantage, especially with the evaluation of the additive parts. Certification is important in the field of MAM in printing parts which requires excellent quality and reliability, such as aerospace industries, where the safety is the main requirement.

6.1.7.1 Proposed solution

MAM part should be certified by setting methods of measurement, monitoring, and control in order to acquire the desirable quality and reliability in parts. Then, the issues in parts will be addressed and analyzed to come up with some actual solutions. Better the measurement and control methods, better the results it gives which accelerates the certification of the MAM method. The measurement method is a trigger to develop standards for addressing these MAM issues and assigning solution to fix them.

6.1.8 Inspection difficulty

Producing complex and non-homogeneous MAM parts, which contains significant numbers of different features, requires high level of inspection. By assigning inspection, the quality of additive part can be retained. The importance of the inspection is equal to the importance of the measurement which has been mentioned in the previous section. Inspection and metallurgical validation methods should be integrated with MAM parts printing. That will allow the production of highly optimized shapes. According to the current study, there is a real need to build non-traditional inspection system inherent to the MAM parts manufacturing [5]. Now, traditional nondestructive inspection methods are used in the MAM method, and these methods are not enough to detect the defect during the printing process.

6.1.8.1 Proposed solution

Studies should be developed on the required inspection methods. Internal inspection needs to be performed during the printing process. It will act like a monitoring system during the process. In other words, layer inspection should be performed inherent to comparison with slandered. In layer inspection, the defects can be detected when it occurs. Let us assume that a very expensive and complex part is being printed. If there are some defects like undesired mechanical properties or porosities in the middle of printing, it would be wise to stop the process and fix the problem rather than rejecting the part after the expensive and time-consuming process is finished.

6.1.9 Scheduling

While increasing the demand and purchasing the MAM parts, the scheduling and nesting processes of metal parts to be printed plays a major role in manufacturing cost. It is beneficial in different ways, such as reducing the operational cost, reducing the parts price, and increasing profit for factories. This issue is also important in supply chain to schedule delivering or requesting metal parts. A very little literature has addressed this issue since it is still new issue [8].

6.1.9.1 Proposed solution

The scheduling issue is very important, especially with the rapid development of MAM method. Existing production planning and scheduling approach should be developed based on preliminary heuristic studies and models generated from this heuristic study. The scheduling should include all the sections of manufacturing system ranging from ordering the raw materials to delivering the metal parts to the customers. Based on that, scheduling system related to all sectors of MAM method with forecasting technologies can be used to organize the production line on MAM with optimal results.

6.1.10 Energy consumption

Energy consumption in MAM method is important to keep the sustainability of the MAM printing. However, it is affecting the microstructure and mechanical properties of the printed parts especially with metal 3D printing process that requires high source of energy, such as EBM process. This issue has a significant influence on the quality of the part.

6.1.10.1 Proposed solution

The relationship between energy consumption and printed metal parts with microstructure and mechanical properties should be analyzed to approach the solution of this problem. This solution can be implemented by using variable optimal energy density for every MAM process that depends on factors, such as part size, part geometry, and material used with strategies to reduce this consumption. Then, the variable optimal energy density map can be generated according to the characteristics of the part, 3D printing machine used, and used material. Therefore, the variable optimal energy used can be specified from the CAD model of the part. This can be done by inserting this CAD model on previously created software responsible for calculating the map of optimal energy density for the 3D printed metal part. This method of using the optimal energy might be useful to remove the undesirable mechanical properties and microstructure from metal printed part.

6.2 MAM defects

The most common challenges and issues that the MAM method faces have been discussed in the previous sections. MAM is a manufacturing process, and every process has its pros and cons. The part produced by MAM also has parameters and defects that may affect the functionality of the part and lead to the failure. These defects are residual stress, fatigue, porosity, variation in mechanical properties, surface oxidation, cracks, delamination, variation in grain size, and heat accumulation. Similarly, parameters such as microstructure, building direction, surface roughness, and powder particle size may lead to these defects or affect the quality and reliability of the printed metal parts. Some of these defects will be discussed next.

6.2.1 Residual stress

Residual stress is a stress stored inside the metallic part in different ways. For example, forging on the surface of some plate creates residual stress which is required to reinforce the plate. Laser peening creates desired compressive residual stress in metal part. Similarly, residual stress is stored in the metal after casting stock material which is mostly undesirable. Therefore, the residual stress can be desirable or undesirable. In the MAM method, one of the main defects is influencing the printed metallic parts with undesirable residual stress which may lead to premature failure. Residual stress in MAM method can lead to distortion failure, changing in tolerance, fracture, fatigue, and delamination. A lot of literature review discussed this problem and it is the focus of leading researchers. Several factors may cause the residual stress in MAM parts, such as rapid heating rate, rapid solidification with high cooling, and melting back previous layer for both PBF and DED MAM categories [11].

6.2.1.1 Proposed solution

It is advantageous to use well maintained heat treatment machine (vacuum furnace) to remove the residual stress from the MAM parts. Based on the author's understanding, melting back of the previous layer is the main issue which generates the residual stress. This melting back can be reduced by reducing the pace of heating while building the next layer. Reducing the melting back can be done by installing a cooling unit to reduce the heat between layers, such as air jet to reduce the heat generated when a new layer is being melted and laying down on the previous one. Also, using variable building direction might help in reducing heat accumulation in one direction. More comprehensive study should be conducted to build inspection system rather than in-situ method. This inspection system is used to analyze the melting back spot and figure out where it exists and assign permanent solution for that. Moreover, post process control (a type of heat treatment control) should be applied on the printed metal parts which contain residual stress after inspection.

6.2.2 Fatigue

Fatigue is a damage in a solid material which causes it to weaken. The main reason for fatigue is fluctuating (cyclic) loads applied on metal parts which causes damage at the end and failure in the parts. Several parameters, such as surface finish, shape and geometry of the part, grain size and direction, functionality of the part, load type, part size, corrosion, temperature, and surface treatment cause fatigue. For the MAM method, the temperature is the most affective parameter to cause fatigue in printed metallic parts. Some literature studies have been performed on this subject, especially on multiaxial fatigue in many industries for small and medium size of metallic parts [1].

6.2.2.1 Proposed solution

Analyze the functionality and the geometry of printed parts to specify where the fatigue can happen for a particular temperature and surface finish. This fatigue could be avoided by assigning specific heat treatment and surface finish. Similarly, the geometry of metallic parts which can induce fatigue can be changed after comprehensive analysis to avoid undesirable residual stress in these features. Finally, using different environments with different temperature during the process can also be used to eliminate fatigue.

6.2.3 Porosity

Another common defect in MAM method that needs to be prevented during the process is porosity. The porosity is a void or hole in the solid printed metal. In the MAM method, the porosity happens between the layers and is mostly inherent to Direct Energy Deposition (DED) category of metal printing, such as WAAM. It occurs while the printed layer is in liquid form because of gas or solidification shrinkage. Porosity is a defect that will end in producing parts that have undesirable mechanical properties, such as low strength and cracks. Unstable deposition process and poor path planning, especially with complex path cause porosity as well [13].

6.2.3.1 Proposed solution

The gap between the torch of DED processes and the melting surface should be continuously adjusted to prevent any slags that might happen and lead to porosity. Using inert vacuum environment during the printing process will eliminate the oxidation that will lead to porosity. Keeping the printing surface (substrate) clean from any dirt or slags either on the previous printed layer or the main surface of the bed will also reduce porosity. While using the optimal gas flow for GT-WAAM process, incorrect gas flow may cause a lot of slags during the printing which may lead to porosity. Optimal energy density can play a main role in eliminating the porosity because it will generate fine printed layer without defects. Creating a map for optimal deposition rate and gap for every specific alloy and reducing the complexity of the printing path can be another solution. Similarly, installing inspection system to check the printing process layer by layer during the process can fix the porosity simultaneously.

6.2.4 Oxidation

Oxidation (Redox) is a process when a chemical reaction happens in metal and leads to change in the structure of atom by losing electrons. Oxidation happens with the availability of oxygen and it is a defect that generates cracks and distortion failure in the printed metal part. It creates a weak insulation material between the printed layers, mostly in DED MAM category. Oxidation is the main reason for variation on mechanical properties. It causes minute or unstable printing gaps between the torch and the bed surface or the previous printed layer. Not enough flow of gas during the process might also lead to oxidation.

6.2.4.1 Proposed solution

The best way to prevent oxidation is by eliminating oxygen from the printing environment. That can be obtained by using air vacuumed chamber 3D metal printer. However, if the vacuumed environment is difficult to be adjusted, optimal stable gap of the torch should be used with optimal gas flow.

6.2.5 Heat accumulation

One of the main defects that interrupt the metal 3D printing part is a heat accumulation. This issue happens when the average temperature of the workpiece increases continuously. That will cause many other defects, such variation in mechanical properties and microstructure evolution, and variation in grain size which will, consequently, lead to undesirable part quality, such structural collapse [14]. This issue is also mostly inherent to DED category of the MAM. According to the author analysis, it happens in DED category mostly because in DED process, metal is melted. As a new layer is created over the previous hot one, it will initiate the heat accumulation. Thus, this issue depends on the philosophy of metal printing process. As compared to PBF, in which lower powered laser is used with lower heat, a new layer of powder will be poured over the previous one and that will absorb the heat of the previous layer.

6.2.5.1 Proposed solution

The previously printed layer should be cooled completely before printing the new layer. That could be performed by installing a cooling unit with the help of in-situ method. The cooling unit will work simultaneously while recording temperature of in-situ method. Similarly, variant building direction will accumulate heat in different direction and reduce the average heat. Heat treatment can be a traditional solution if the workpiece could not be collapsed during the printing process. According to the author's understanding, another way to solve the issue of heat accumulation or residual stress is by using nucleation after printing. Nucleation is a process of formation of a new crystalline structure during solidification. This process might be applied on the printed parts by heating it to a temperature lower than the melting point and building new crystalline structure with the desired mechanical properties, grain size, free residual stress and heat accumulation.

6.2.6 Cracks and delamination

Two types of cracks mainly occur in MAM method (DED processes) and they are: grain boundary cracks and solidification cracks. Cracks happen because of the method of solidification. Delamination is a type of failure in metal alloys or composite material. In defective parts, the layers of metal are separated because of repeated fluctuated stress and the material characteristic of the deposited metal. A drawback of delamination is that it cannot be repaired by post-process heat treatment, but it could be avoided [13].

6.2.6.1 Proposed solution

The only way to fix the problem of delamination and cracks is by preventing them. It can be achieved by optimizing solidification. Preheating the printing environment and substrate can be a good solution as well. Avoid using combined material in the DED processes that might separate the layers between them when the dissimilar material has significant difference in the chemical reaction and solubility. Comprehensive study should be performed on the metals which are difficult to melt and solidify without cracks and delamination such as Inconel. Developing temperature map of the printed surface, that might help to optimize the preheating, can avoid cracking and delamination. The high cooling rate is also a cause of cracks and delamination. However, it is mentioned previously that it is a good solution to avoid some defect such as heat accumulation. Therefore, an optimal cooling rate should be used to avoid and consider these defects.

7. Summary and conclusion

In this chapter, a review on design for additive manufacturing was conducted to address and answer some concerns about Metal Additive Manufacturing (MAM). These concerns were related to challenges, issues, and defects which are prevalent in MAM method in factories while printing small and medium sized parts. Explanations about the printing procedures and materials are also discussed in this chapter. Moreover, potential approach and solutions are inferred by the author to diminish these problems. Comprehensive literature review was developed from a recent research paper to address and understand these manufacturing obstacles. The MAM method is divided into two main processes: PBF and DED. These processes were explained in detail by the author.

The main obstacles which the MAM methods encounter are categorized into two types: challenges (difficulties) and defects. Several obstacles have been explained in this chapter, such as repeatability, material issue, schedule, certification, heat treatment, and inspection. Also, several defects which obstruct high quality and reliability, such as fatigue, stress analysis, porosity, heat accumulation, delaminating, and cracks were listed and discussed. For every challenge and defect, a potential solution was suggested and discussed by the author. Some of these suggested solutions need to be tested and validated on actual metal three-dimensional printing processes. Other suggested solutions, such as inspection and orientation, should be implemented by visiting manufacturing companies and making local study on to solve the problems.

However, based on the literature review and analysis, DED processes, especially WAAM, have more defects and challenges than PBF processes. Some of these challenges and defects are common in three-dimensional printing, a fast-growing manufacturing process. More study and analysis are required to fix these obstacles and make MAM method smooth. Such concurrent failures can induce huge losses in industries and force them out of business. Therefore, future studies can eliminate these failures and yield higher production rate and profits.

Acknowledgements

The authors wish to thank the resources of the Department of Industrial and Management Systems Engineering, West Virginia University.

Author details

Omar Ahmed Al-Shebeeb
Department of Industrial and Management Systems Engineering, West Virginia University, Morgantown, WV, USA

*Address all correspondence to: oaalshebeeb@mix.wvu.edu

References

[1] Molaei R, Fatemi A. Fatigue design with additive manufactured metals: Issues to consider and perspective for future research. Procedia Engineering. 2018;**213**:5-16

[2] Challenges Associated With Additive Manufacturing. Available from: https://www.forbes.com/sites/forbestechcouncil/2018/03/28/challenges-associated-with-additive-manufacturing/#434a23f66db0

[3] Wohlers T, Gornet T. History of additive manufacturing. Wohlers Report. 2014;**24**:118

[4] Shvaiko P, Euzenat J. Ontology matching: State of the art and future challenges. IEEE Transactions on Knowledge and Data Engineering. 2013;**25**(1):158-176

[5] Additive Manufacturing and Inspection Difficulties. Available from: https://www.qualitymag.com/articles/94838-additive-manufacturing-and-inspection-difficulties

[6] Challenges Associated With Additive Manufacturing. Available from: https://www.forbes.com/sites/forbestechcouncil/2018/03/28/challenges-associated-with-additive-manufacturing/#1d2dfe376db0

[7] Graziosi S, Rosa F, Casati R, Solarino P, Vedani M, Bordegoni M. Designing for metal additive manufacturing: A case study in the professional sports equipment field. Procedia Manufacturing. 2017;**11**:1544-1551

[8] Chergui A, Hadj-Hamou K, Vignat F. Production scheduling and nesting in additive manufacturing. Computers and Industrial Engineering. 2018;**126**:292-301

[9] Richardson BS, Lind RF, Lloyd PD, Noakes MW, Love LJ, Post BK. The design of an additive manufactured dual arm manipulator system. Additive Manufacturing. 2018;**24**:467-478

[10] Solutions for Additive Manufacturing and Powder Injection Molding. Available from: https://www.verder-scientific.com/powder-metallurgy/?gclid=CjwKCAiAiuTfBRAaEiwA4itUqF5IjBOWTPX8OANgpT_5ERqAqMH7vmqNS-K3owJFGBH1DPRSfGLLrBoCj_QQAvD_BwE

[11] Li C, Liu ZY, Fang XY, Guo YB. Residual stress in metal additive manufacturing. Procedia CIRP. 2018;**71**:348-353

[12] Liu ZY, Li C, Fang XY, Guo YB. Energy consumption in additive manufacturing of metal parts. Procedia Manufacturing. 2018;**26**:834-845

[13] Wu B et al. A review of the wire arc additive manufacturing of metals: Properties, defects and quality improvement. Journal of Manufacturing Processes. 2018;**35**:127-139

[14] Montevecchi F, Venturini G, Grossi N, Scippa A, Campatelli G. Heat accumulation prevention in wire-arc-additive-manufacturing using air jet impingement. Manufacturing Letters. 2018;**17**:14-18

[15] Wu B, Pan Z, Ding D, Cuiuri D, Li H. Effects of heat accumulation on microstructure and mechanical properties of Ti6Al4V alloy deposited by wire arc additive manufacturing. Additive Manufacturing. 2018;**23**:151-160

[16] Li X, Tan W. Numerical investigation of effects of nucleation mechanisms on grain structure in metal additive manufacturing. Computational Materials Science. 2018;**153**:159-169

[17] Du J, Wang X, Bai H, Zhao G, Zhang Y. Numerical analysis of fused-coating metal additive manufacturing. International Journal of Thermal Sciences. 2017;**114**:342-351

[18] Meier C, Weissbach R, Weinberg J, Wall WA, Hart AJ. Critical Influences of Particle Size and Adhesion on the Powder Layer Uniformity in Metal Additive Manufacturing. arXiv Prepr. arXiv1804.06822; 2018

[19] Murr LE, Johnson WL. 3D metal droplet printing development and advanced materials additive manufacturing. Journal of Materials Research and Technology. 2017;**6**(1):77-89

[20] Rojas JGM, Wolfe T, Fleck BA, Qureshi AJ. Plasma transferred arc additive manufacturing of nickel metal matrix composites. Manufacturing Letters. 2018;**18**:31-34

[21] Yoozbashizadeh M, Yavari P, Khoshnevis B. Novel method for additive manufacturing of metal-matrix composite by thermal decomposition of salts. Additive Manufacturing. 2018;**24**:173-182

[22] Livescu V, Knapp CM, Gray GT III, Martinez RM, Morrow BM, Ndefru BG. Additively manufactured tantalum microstructures. Materialia. 2018;**1**:15-24

[23] Paris H, Mandil G, et al. The development of a strategy for direct part reuse using additive and subtractive manufacturing technologies. Additive Manufacturing. 2018;**22**:687-699

[24] Vayre B, Vignat F, Villeneuve F. Metallic additive manufacturing: State-of-the-art review and prospects. Mechanics & Industry. 2012;**13**(2):89-96

[25] Ma F, Zhang H, Hon KKB, Gong Q. An optimization approach of selective laser sintering considering energy consumption and material cost. Journal of Cleaner Production. 2018;**199**:529-537

[26] Marrey M, Malekipour E, El-Mounayri H, Faierson EJ. A framework for optimizing process parameters in powder bed fusion (pbf) process using artificial neural network (ann). Procedia Manufacturing. 2019;**34**:505-515

[27] Greiner S, Wudy K, Wörz A, Drummer D. Thermographic investigation of laser-induced temperature fields in selective laser beam melting of polymers. Optics and Laser Technology. 2019;**109**:569-576

[28] Williams SW, Martina F, Addison AC, Ding J, Pardal G, Colegrove P. Wire+ arc additive manufacturing. Materials Science and Technology. 2016;**32**(7):641-647

[29] Simar A, Godet S, Watkins TR. Highlights of the special issue on metal additive manufacturing. Materials Characterization. September 2018;**143**:5-17

[30] Stock Powders. Available from: https://www.i3dmfg.com/?_vsrefdom=adwords&gclid=Cj0KCQiA_4jgBRDhARIsADezXciAgnquGbB7kgwCKgLxH9CSy9wyLHHjLNZCBxph7EerloMPUgIqj3QaAiyhEALw_wcB

[31] Zhao Y et al. The stress coupling mechanism of laser additive and milling subtractive for FeCr alloy made by additive--subtractive composite manufacturing. Journal of Alloys and Compounds. 2018;**769**:898-905